Industry 5.0 for Society 5.0: Advancing Medical IoT and Smart Healthcare

(Part 1)

Edited by

Parikshit N. Mahalle
Department of AI & DS
Vishwakarma Institute of Technology
Pune, India

Gitanjali R. Shinde
Department of CSE AIML
Vishwakarma Institute of Technology
Pune, India

Namrata N. Wasatkar
Department of Computer Engineering
Vishwakarma Institute of Technology
Pune, India

&

Prashant R. Anerao
Department of Mechanical Engineering
Vishwakarma Institute of Technology
Pune, India

Industry 5.0 for Society 5.0: Advancing Medical IoT and Smart Healthcare *(Part 1)*

Editors: Parikshit N. Mahalle, Gitanjali R. Shinde, Namrata N. Wasatkar & Prashant R. Anerao

ISBN (Online): 979-8-89881-045-0

ISBN (Print): 979-8-89881-046-7

ISBN (Paperback): 979-8-89881-047-4

Published by Bentham Science Publishers Pte. Ltd. Singapore,

in collaboration with Eureka Conferences, USA. All Rights Reserved.

First published in 2025.

need for a court order if at any point you breach any terms of this License Agreement. In no event will any delay or failure by Bentham Science Publishers in enforcing your compliance with this License Agreement constitute a waiver of any of its rights.

3. You acknowledge that you have read this License Agreement, and agree to be bound by its terms and conditions. To the extent that any other terms and conditions presented on any website of Bentham Science Publishers conflict with, or are inconsistent with, the terms and conditions set out in this License Agreement, you acknowledge that the terms and conditions set out in this License Agreement shall prevail.

Bentham Science Publishers Pte. Ltd.
No. 9 Raffles Place
Office No. 26-01
Singapore 048619
Singapore
Email: subscriptions@benthamscience.net

CONTENTS

FOREWORD

I am extremely happy to write the Foreword for an edited book titled "Industry 5.0 for Society 5.0: Advancing Medical IoT and Smart Healthcare" by prestigious Bentham Science Press. This book provides readers an opportunity to explore the transformative potential of Industry 5.0 and its profound impact on society. It is an opportunity to delve into the convergence of advanced technologies and human-centric solutions, shaping a new era of industrial and societal development.

Unlike its predecessor, Industry 4.0, which focused on data digitization and machine automation, Industry 5.0 aims to rebuild a human-centric environment where humans and machines work together to boost work efficiency. This shift towards a more collaborative and cooperative approach between humans and machines is pivotal in fostering a smart manufacturing system.

The book also delves into the vision of Industry 5.0 leading toward Society 5.0, envisioning a smart society equipped with new versions of education and educators 5.0. This evolution aims to lead to a society where humans are equipped with smart collaborative and cooperating robots, emphasizing the transformative potential of these advancements. The book acknowledges that despite Industry 5.0's potential contributions to the world economy, ecology, and society, there will be challenges during its implementation. It also highlights the need for research in prospective areas to address these challenges and maximize the benefits of Industry 5.0 for society. The book includes key areas like medical IoT and healthcare applications.

I would like to congratulate all editors for coming up with valuable contributions on very apt topics and I am sure that this book will be well received by all the stakeholders.

Varsha Hemant Patil
Ex- Chairman
Department of Computer Engineering
Savitribai Phule Pune University
Pune, Maharashtra, India

&

Vice Principal
Department of Computer Engineering
Matoshri College of Engineering and Research Center
Maharashtra, India

PREFACE

Industry 5.0 represents a significant evolution in the industrial landscape, focusing on the collaboration between humans and advanced technologies, such as artificial intelligence (AI) and the Internet of Things (IoT). This paradigm shift aims to create a more human-centric approach to industry, enhancing productivity while prioritizing the well-being of individuals and society. Society 5.0, on the other hand, is envisioned as a super-smart society that leverages technology to address social challenges and improve quality of life. It builds upon the foundations of Society 4.0, aiming to create a prosperous, human-centered environment where technology serves to solve pressing societal issues. Society 5.0 envisions a society where advanced IT technologies, the Internet of Things, robots, artificial intelligence, and augmented reality are actively used in everyday life, industry, healthcare, and other spheres for economic advantage and the benefit of citizens. It aims to create a human-centric super smart society with high-quality, comfortable lives.

The relationship between Industry 5.0 and Society 5.0 is described as reciprocal, emphasizing the need to maintain and develop the relationship between industry and society. This reciprocal relationship is crucial for societal development and the integration of advanced technologies to improve welfare. Both Industry 5.0 and Society 5.0 emphasize human-centric approaches, focusing on the integration of technology with the human factor for proper management and achieving the best results. The convergence of Industry 5.0 and Society 5.0 is paving the way for a revolutionary approach to healthcare through the advancement of Medical IoT and smart healthcare solutions. By focusing on human-centric designs and leveraging cutting-edge technologies, the healthcare sector can significantly improve patient outcomes and operational efficiency, ultimately contributing to a healthier society.

This book is a one stop shop that offers the readers everything he/she needs to know or use industry 5.0 for society 5.0. evolution towards industry 5.0 which includes Industry 5.0 for Society 5.0, enabling technologies, opportunities, challenges and future perspectives. The book offers a basic understanding of medical IoT and healthcare applications, and the integration of Medical IoT within the framework of Industry 5.0, patient-centered care, IoT applications and ethical data practices in digital health. Techniques and case studies that include smart computing on the convergence of Industry 5.0 for Society 5.0 for medical IoT and smart healthcare are introduced to the reader. An outlook from where the readers can build upon and work towards developing their applications is presented in the book.

This book provides an overview of various use cases which can be build upon the convergence of Industry 5.0 for Society 5.0 and mainly the role of Industry 5.0 in above mentioned subject areas. A few key features of this books are as follows:

• Discusses the broad background of Industry 5.0, Society 5.0 & its fundamentals.

• The role of Industry 5.0 in medical IoT and healthcare applications and various use cases towards human-centric computing.

• Discusses various technologies, methodologies and approaches that play a prominent role in medical IoT, smart healthcare and Industry 5.0.

• The role of Industry 5.0 for preventive and sustainable healthcare.

In a nutshell, this book displays all information (basic and advanced) that a novice and advanced reader needs to know regarding the role of Industry 5.0 for Society 5.0. The book also motivates the use of appropriate technology for the better development of applications. The book also contributes to social responsibilities by laying down the foundation for the development of applications that can help in making day to day activities easier by meeting the requirements of important sector of healthcare and other vital aspects of human lives.

Parikshit N. Mahalle
Department of AI & DS
Vishwakarma Institute of Technology
Pune, India

Gitanjali R. Shinde
Department of CSE AIML
Vishwakarma Institute of Technology
Pune, India

Namrata N. Wasatkar
Department of Computer Engineering
Vishwakarma Institute of Technology
Pune, India

&

Prashant R. Anerao
Department of Mechanical Engineering
Vishwakarma Institute of Technology
Pune, India

List of Contributors

Aniket S. Ingavale	School of Engineering and Technology, DES Pune University, Pune, India
B. Subashini	Department of Mechatronics Engineering, Kumaraguru College of Technology, Coimbatore, Tamil Nadu, India
Grishma Y. Bobhate	Department of Computer Science & Engineering-Artificial Intelligence & Machine Learning, Vishwakarma Institute of Technology, Pune, India
Jayashri Bagade	Department of Information Technology, Vishwakarma Institute of Technology, Pune, India
K. Swetha	Department of Computer Science and Engineering, Kumaraguru College of Technology, Coimbatore, Tamil Nadu, India
L. Feroz Ali	Department of Mechatronics Engineering, Sri Krishna College of Engineering and Technology, Coimbatore, Tamil Nadu, India
Mrunmayee Solkar	Department of Information Technology, Vishwakarma Institute of Technology, Pune, India
Nilesh Sable	Department of Information Technology, Vishwakarma Institute of Technology, Pune, India
Pranjal S. Pandit	Department of Computer Science & Engineering- Artificial Intelligence, Vishwakarma Institute of Technology, Pune, India
Pallavi Devendra Deshpande	Department of Electronics and Telecommunication Engineering, Vishwakarma Institute of Technology, Pune, India
P. Sudam Sekhar	Department of Mathematics and Statistics, Vignan University, Guntur, India
R. Raffik	Department of Mechatronics Engineering, Kumaraguru College of Technology Coimbatore, Tamil Nadu, India
R.P. Roshan	Department of Mechatronics Engineering, Kumaraguru College of Technology Coimbatore, Tamil Nadu, India
R. Asvitha	Department of Mechatronics Engineering, Kumaraguru College of Technology, Coimbatore, Tamil Nadu, India
Rutuja Diwate	Department of Information Technology, Vishwakarma Institute of Technology, Pune, India
Rohini Chavan	E & TC Department, Vishwakarme Institute Of Technology, Pune, India
Rakesh Nayak	Department of CSE, OP Jindal University, Raigarh, India
S. Karthikeyan	Department of Mechanical Engineering, Christian College of Engineering and Technology, Oddanchatram, Tamil Nadu, India
Saurabh Sathe	Department of Computer Science, San Jose State University, San Jose, California, USA
Santosh Kumar	Department of Artificial Intelligence & Data Science, Vishwakarma Institute of Technology, Pune, Maharashtra 411048, India
Soumya Ranjan Mahanta	Department of Computer Science, Utkal University, Bhubaneswar, Odisha, India
Shreyash Shabadi	E & TC Department, Vishwakarme Institute Of Technology, Pune, India

Tanmay Anil Rathi   IT Department, New York University, New York, NY 10012, USA

Umashankar Ghugar   Department of CSE, University Institute of Engineering (UIE), Chandigarh University, Mohali, Punjab, India

W. Aagasha Maria   Department of Mechatronics Engineering, Kumaraguru College of Technology, Coimbatore, Tamil Nadu, India

CHAPTER 1

Industry 5.0 for Society 5.0 - Enabling Technologies, Opportunities, Challenges and Future Perspectives

R. Raffik[1,*], **S. Karthikeyan**[2], **R.P. Roshan**[1] and **K. Swetha**[3]

¹ Department of Mechatronics Engineering, Kumaraguru College of Technology Coimbatore, Tamil Nadu, India

² Department of Mechanical Engineering, Christian College of Engineering and Technology, Oddanchatram, Tamil Nadu, India

³ Department of Computer Science and Engineering, Kumaraguru College of Technology, Coimbatore, Tamil Nadu, India

Abstract: The industry 5.0 (I5.0) paradigm shift aims to advance human-centricity, resolving problems, and the ability to make decisions using revolutionary technologies. Technology is a tool that humans may use to enhance their skills and produce greater results, not to replace them. Society 5.0 (S5.0), conversely coined as "Super Smart Society," was conceived by the government of Japan. Human-centered societies strike a balance between economic development and socio-problem resolution by offering products and services. The chapter clarifies how Industry 5.0 and Society 5.0 working together may create a peaceful coexistence that promotes economic progress, innovation, and social well-being. Technological transitions and developments like the Industrial Internet of Things (IIoT), Advanced Robotics, Cobots, Artificial Intelligence (AI), Cybersecurity, and Big data, Cloud computing, enhance the lifestyle of human life and the performance of industries through home automation and industrial automation. The study elaborates on significant areas for technology collaborations to make Industrial 5.0 and Society 5.0 combination to revolutionize various sectors such as inclusive education, health care, power plants, smart cities, sustainable development, and collaborative robots. Industry 5.0 and Society 5.0 show innovation-driven good social development in future generations. The present research adds to the current conversation on the incorporation of new technologies by offering insights into the cooperative development of companies and communities in the direction of a resilient and human-centered future. This chapter also discusses the futuristic perceptions and relevant challenges in attaining Industry 5.0 and Society 5.0.

---

* **Corresponding author R. Raffik:** Department of Mechatronics Engineering, Kumaraguru College of Technology Coimbatore, Tamil Nadu, India; E-mail: raffik.r.mce@kct.ac.in

Parikshit N. Mahalle, Gitanjali R. Shinde, Namrata N. Wasatkar & Prashant R. Anerao (Eds.)

Keywords: Artificial intelligence, Industrial internet of things, Industry 5.0, Industrial automation, Smart cities, Society 5.0, Sustainable development.

INTRODUCTION

The fifth industrial revolution is a modern approach to manufacturing that aims to create a system that is more centered around humans, sustainable, and resilient [1]. The I5.0 paradigm emphasizes how crucial it is to give immense importance to this matter of agility and resiliency in systems, and it accomplishes this by using adaptable and flexible technologies. Moreover, the I5.0 approach seeks to promote sustainability by respecting the boundaries of our planet and encouraging diversity, empowerment, and talent. I5.0 has a more comprehensive perspective on the function of the industry, extending beyond just creating jobs and economic growth. The aim is to establish an industry that is both durable and sustainable, and that considers the planet's constraints. Additionally, the well-being of industrial workers is given top priority [2]. The concept of I5.0 addresses the concerns; the emphasis of Industry 4.0 is on the implementation of digitization and AI-powered technological advances have overshadowed the fundamental ideas of fairness and ecological balance. Emphasizing the importance of investigation and industry innovation, I5.0 provides a distinctive approach. It aims to use these tools to deliver long-term benefits to humanity within the limits of our planet [3].

The Japanese government introduced S5.0 in January 2016 - a concept of a futuristic society driven by technological and scientific advancements [4]. The aim of this vision is to create a society that is highly intelligent and human-centered, with a focus on meeting individual needs. The goal of S5.0 is to provide a comfortable and fulfilling life for everyone by merging the physical and cyberspace realms, using Advanced technologies, including 5G, AI, and Big Data, which are currently being utilized [5]. By leveraging these technologies, S5.0 intends to empower individuals with tailored goods and services that cater to their unique needs. To gain a deeper comprehension of the correlation and integration linking I5.0 and S5.0, it is necessary to make a comparison based on their respective definitions. Four dimensions - goal, value, organization, and technology - can be used to compare these two concepts by visualizing their similarities and differences - have been used. The comparison diagram serves to highlight both the differentiation between and resemblance between I5.0 and S5.0, as depicted in Fig. (**1**), providing valuable insights for further exploration and development.

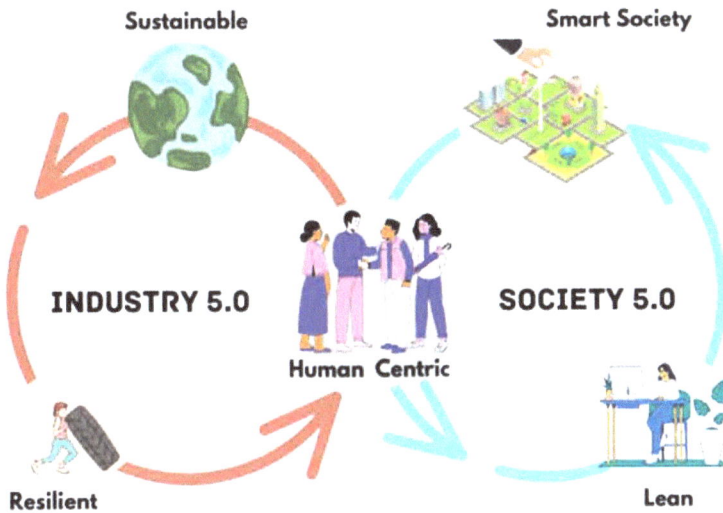

Fig. (1). Comparison of I5.0 & S5.0.

I5.0 and S5.0 are two concepts that prioritize human centricity, which means putting humans at the center of their development and implementation. I5.0 aims to foster human creativity in industries and then promote a shift towards a more human-centered, resilient, and sustainable future. It seeks to address current issues with industrialization and living standards by prioritizing the needs and well-being of people [6]. S5.0, on the other hand, aims to build a lean, human-centered, and superintelligent civilization providing a comfortable additionally good future shared by all. It focuses on developing technology and infrastructure that can enhance the quality of life for people, while also promoting social equity and inclusivity. In the future, individual vitality and demand will play a crucial role in both industry and society, and these concepts will continue to shape life and work.

In I5.0, the product lifecycle has evolved to encompass various aspects such as research and development, production efficiency, tailored services, and recycling. There is a growing trend towards developing value-creation processes at the later stages of the product lifecycle, which is a positive step towards sustainable and responsible manufacturing. S5.0, on the contrary, is expected to be an inclusive and all-encompassing framework that will generate value by means of personalized services, advanced transportation systems, and smart manufacturing systems. This will undoubtedly lead to a more efficient, productive, and environmentally conscious society. The manufacturing cell, factory, and supply chain are the backbone of the industry. I5.0 highlights the need for industrial resilience during COVID-19. S5.0 envisions merging the digital and physical worlds to address social problems. Industry plays a vital role in society's organization [7]. The latest technological advancements, such as advanced wireless networks, big data, AI, and digital twins, among others, are having a significant impact on I5.0 and S5.0. These innovative technologies are spearheading the intellectualization and digitization movement, they will open the door to a new era in various industries and even the entire society. The ability of emerging innovations to change how people live, and work is immense and presents exciting opportunities for progress and growth.

The evolution from Industry 4.0 to Industry 5.0 represents a significant paradigm shift in the way we approach manufacturing, technology, and society. Industry 5.0 emphasizes the importance of human-centricity, sustainability, and resilience, aiming to create systems that are not only efficient but also equitable and environmentally conscious. This shift is driven by the need to address the limitations of Industry 4.0, where the focus on digitization and automation often overshadowed the fundamental aspects of human well-being and ecological balance.

Similarly, the concept of Society 5.0, introduced by the Japanese government, envisions a "Super Smart Society" where technological advancements are harnessed to create a highly intelligent, human-centered society. The goal of Society 5.0 is to merge the physical and cyberspace realms to meet individual needs and improve the quality of life, promoting social equity and inclusivity.

Motivation of this Chapter

The motivation behind exploring the integration of Industry 5.0 and Society 5.0 lies in the potential synergies that can be harnessed to create a more sustainable, resilient, and equitable future. By combining the strengths of both paradigms, we can address the current challenges faced by industries and societies, such as economic disparities, environmental degradation, and social inequities.

Human-Centric Approach:
- *Enhancing Human Skills:* Unlike the traditional view where technology replaces human roles, Industry 5.0 sees technology as a tool to enhance human skills and creativity. This approach fosters a collaborative environment where humans and machines work together, leading to more innovative and personalized solutions.
- *Improving Quality of Life:* Society 5.0 aims to create a society where technology meets individual needs, enhancing overall well-being. This includes providing tailored goods and services, improving healthcare, and creating smart cities that offer a higher quality of life.

Sustainability and Resilience:

- *Environmental Consciousness*: Industry 5.0 emphasizes sustainable practices, respecting planetary boundaries, and promoting the circular economy. This focus ensures that industrial growth does not come at the expense of the environment.
- *Social Equity and Inclusivity*: Society 5.0 strives to create an inclusive society where everyone benefits from technological advancements. This includes addressing social challenges, promoting equity, and ensuring that no one is left behind in the technological revolution.

Technological Advancements:

- *Advanced Technologies*: The integration of advanced technologies such as AI, IIoT, robotics, big data, and cloud computing can revolutionize various sectors, enhancing efficiency, productivity, and innovation.
- *Interconnected Systems*: The use of technologies like the Internet of Things (IoT) and digital twins allows for the creation of interconnected systems that can monitor and optimize processes in real time, leading to smarter and more responsive industrial and social infrastructures.

Economic Growth and Innovation:

- *Boosting Productivity:* By leveraging technologies that improve automation, decision-making, and process optimization, industries can achieve higher productivity levels and economic growth.
- *Fostering Innovation:* The collaborative nature of Industry 5.0 encourages continuous exploration and innovation, driving the development of new products, services, and business models that cater to the evolving needs of society.

Addressing Challenges:

- *Cybersecurity and Job Displacement:* As we transition to Industry 5.0, it is crucial to address the potential negative aspects such as cybersecurity risks and job displacement. Ensuring a fair and responsible implementation of these technologies is essential to mitigate these challenges.
- *Social Challenges:* Society 5.0 aims to use technology to solve social problems, from healthcare and education to transportation and urban development, creating a more cohesive and supportive community.

By examining the interplay between Industry 5.0 and Society 5.0, this chapter aims to provide valuable insights into how these paradigms can work together to shape a more sustainable, resilient, and human-centered future. Through collaborative efforts and continuous innovation, we can harness the transformative power of technology to create a world that supports economic prosperity, social well-being, and environmental sustainability.

Contribution of this Chapter

This chapter explores the combination of I5.0 and S5.0, clarifying their interdynamics and their essential role in enabling technologies. I5.0 encourages human-machine cooperation in the manufacturing sector, utilizing technologies like collaborative robotics and inventory management to improve productivity and safety. On the other hand, S5.0 utilizes AI, blockchain, and sustainable development to address social challenges and encourage inclusive growth. By examining the reciprocal impacts between these two paradigms, this chapter highlights the potential synergies and transformative power of technology in shaping the future with artificial intelligence. However, it stresses the need for collaborative efforts across various fields to ensure that technology serves the best interests of society, supporting economic prosperity while prioritizing social and environmental sustainability. Through continuous exploration and innovation, I5.0 and S5.0 can work together to create a more equitable and prosperous world.

Structure of this Chapter

This chapter investigates the potential synergies between I5.0 and S5.0 to enhance society and economy. Section 1 discusses how humans can profit from modern technology. What is the importance of artificial intelligence (AI) with respect to I5.0. The use of AI by businesses to automate processes, enhance decision-making, and boost productivity is covered in Section 2. Section 3 discusses how artificial intelligence (AI) might improve supply chains' efficiency, lower waste, and help with inventory management. In I5.0, cloud computing is essential because it streamlines the utilization of computer resources and reduces costs for businesses, as discussed in Section 4. Section 5 discusses how industrial machines and gadgets are interacting with one another considering the Internet of Things (IoT). Using this technology, businesses may reduce downtime, remotely monitor processes, and make more informed decisions. However, there are several negative aspects of I5.0, including job displacement and cybersecurity risks. In Section 6, we discuss how to resolve these problems and ensure a fair and responsible implementation of I5.0.

I5.0 AND S5.0 WITH AI ENHANCEMENTS

Artificial intelligence (AI) can be defined as the intelligence that computers gain from algorithms that are given to them to solve issues, make decisions, and execute tasks that humans would typically accomplish. In other words, artificial intelligence (AI) models human cognition and behavior in computers. I4.0 artificial intelligence (AI) has shown immense potential in supply chain management and robotics manufacturing, vision for machines, analytics for prediction, inventory management, and predictive maintenance. Given the rise of

4.0 Industry, I5.0, and the progressive notion of S5.0, it is imperative to investigate the consequences, prospects, and challenges presented by sophisticated generating intelligent (AI) systems, particularly ChatGPT and its equivalents [8]. A new paradigm known as "I5.0" emphasizes human-robot cooperation [9]. In the future, according to S5.0, technology will improve people's capacity to deal with societal issues. Nevertheless, it is impossible to ignore the ethical and societal issues that S5.0 brings. Making sure that the advantages of AI are distributed fairly, especially to underserved populations, becomes a top priority. In the framework of S5.0, this study aims to investigate these issues and provide insight into how ChatGPT might be used as a socially beneficial tool. Co-creation is where generative AI in I5.0 shines. For engineers, designers, and construction workers, ChatGPT serves as a collaborative collaborator, supporting ideation sessions and producing design concepts based on human input. Artificial intelligence's application to urban planning is a vital aspect of S5.0. In S5.0, environmental sustainability is becoming increasingly important [10]. Eco-friendly construction techniques are developed with the use of generative AI technologies. ChatGPT creates sustainable building designs and encourages green construction practices by examining building materials, energy usage trends, and construction procedures. Construction industry experts benefit from data-driven insights provided by generative AI, which facilitates difficult decision-making. ChatGPT can produce comprehensive reports that highlight hazards, chances for cost savings, and environmental effects related to various building techniques and materials by analyzing large datasets. With their capacity for collaborative problem-solving, real-time data analysis, and predictive modeling, those intelligent systems have completely transformed the industry. The industry's emphasis on human welfare is aligned with AI-driven safety measures and real-time monitoring systems that prioritize the health and security of workers [11].

The construction sector is evolving across sectors 4.0, 5.0, and 5.0 of society thanks to ChatGPT and related AI. They are positioned as the pillars of a creative, inclusive, and resilient future for both society and the industry because of their capacity to improve communication, collaborate with others, and spread knowledge. By responsibly utilizing these technologies, we can expand the realm of what is feasible and make the world an improved and more connected place for all [12]. AI and edge computing involve two creative innovations that are driving I5.0. Edge computing, which processes data getting nearer to the information sources to decrease delay as well as improve real-time responsiveness, is a perfect fit for I5.0's dynamic requirements [13]. A subject of considerable attention and inquiry is the potential cooperative effect of modern computers and computational intelligence as I5.0 continues to change manufacturing, supply chain administration, and several additional industrial areas. Fig. (**2**) illustrates how edge computing for industrial systems provides instantaneous insights, decreased latency, and enhanced autonomy. The concept of decentralized information handling at the periphery of the network forms its basis. Though there are several theoretical frameworks for I5.0, there are few practical instances illustrating how these technologies will interact to shape its development. To bridge this gap, this study undertakes an empirical appraisal to examine the benefits and challenges of merging AI with edge computing in an actual industrial setting [14].

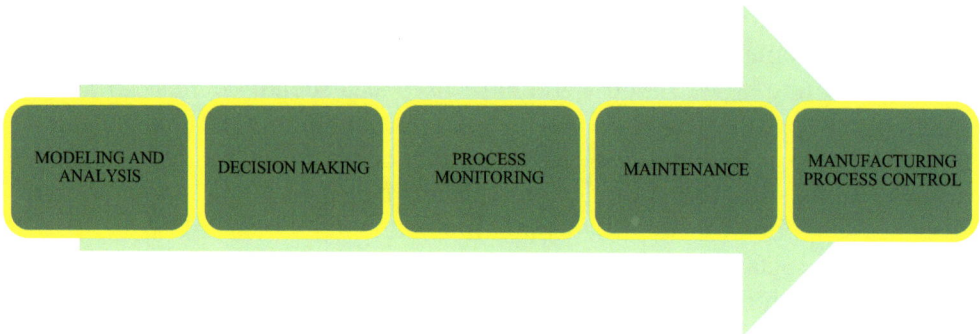

Fig. (2). The integration of AI in I5.0 & S5.0.

AI IMPROVES INVENTORY MANAGEMENT

Because AI can analyze massive amounts of data and produce insights that can be put into practice, its application in the management of supply chains has grown in popularity. In today's business environment, inventory management is essential to both operational effectiveness and customer satisfaction [15]. Inventory management systems have benefited from the integration of AI. Artificial Intelligence has transformed conventional inventory management methods through the analysis of large datasets and trend prediction. A Machine learning algorithm can also improve inventory control by predicting demand more precisely and identifying trends and abnormalities through AI-driven analytics to optimize different supply chain elements. Fig. (3) indicates the ecosystem of cloud manufacturing in inventory management.

Efficient inventory management goes beyond merely storing and transporting products. It entails managing a vast amount of data to optimize time, money, and labour. Failure to recognize and respond to shifting demands is the most common cause of overstocking and understocking, necessitating skilled analysts and business modeling professionals to anticipate these changes. Managers and analysts must adapt to each product niche, considering multiple warehouses and location-specific demands. Artificial Intelligence is a valuable tool in addressing these challenges. Businesses can control inventories and keep an eye on operations by utilizing predictive AI inventory models. AI's ability to analyze over 50 variables is essential for effective scheduling, inventory control, and delivery planning.

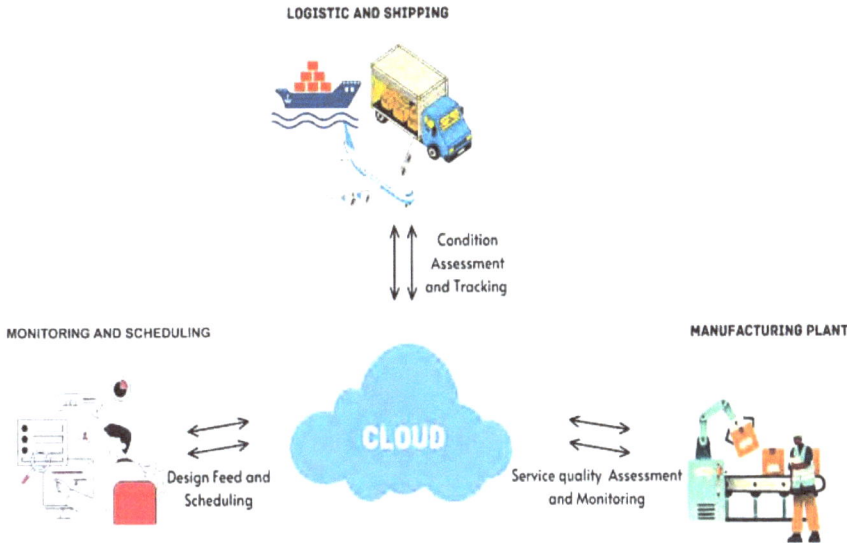

Fig. (3). The ecosystem of cloud manufacturing.

Innovative AI technology can efficiently process data to assist businesses in adapting and responding to situations in significant ways. For example, if a local hockey team advances in the Stanley Cup, there will be a surge in demand for snacks and beverages on the next match day in that region. AI can analyze this information and suggest overstocking in that area. Ensuring client pleasure and a sense of fulfillment requires effective inventory management and fulfillment. In any warehouse, problems with scheduling and inventory control can cause shortages and delays, which affect earnings. Artificial Intelligence (AI) can help with accurate inventory planning by analyzing customer behaviour patterns and other factors. Furthermore, by recommending the best routes, an AI with proper training can increase delivery efficiency and automate the stocking process. Businesses may respond to client demand more quickly by reducing the risk of stocking errors through effective AI inventory management [16]. Furthermore, AI can help businesses establish effective transportation from factories to warehouses by utilizing insights obtained through data mining. This is especially crucial for perishable products with shorter shelf lives.

For quite some time now, robots have been a fixture in the market. Companies like Amazon have already integrated robots into their everyday organization, and there are numerous advantages that make robots preferable to human employees, particularly in routine operations. Robots can tirelessly transport products throughout warehouses, locate items, and scan their conditions. Machines can operate 24/7, with more efficient time management for each task. This can significantly reduce expenses, freeing up staff to focus on more critical and time-sensitive activities that require human insight. Artificial Intelligence (AI) can further enhance the process. When combined with intelligent data analysis, demand trend forecasting, and optimal delivery route recommendations, it becomes a powerful tool capable of fully automating internal warehouse operations.

AI-ENHANCED IIOT

The emergence of interconnected devices has led to an abundance of data that can provide valuable insights. Within the industrial sector, IIoT datasets are particularly advantageous. It is important to note that IIoT datasets differ significantly from regular IoT datasets, as shown in Fig. (4), and are acquired under more rigorous circumstances [17]. The implementation of Industrial Automation Tools and the Industrial Internet of Things (IIoT) is needed in supply chain control and inventory tracking systems. The integration of IIoT with industrial automation enhances real-time tracking, improves efficiency, and reduces errors in the supply chain, aligning with the goals of Industry 5.0 by promoting more resilient and flexible industrial processes [18]. Various technologies and methodologies facilitate seamless interaction between humans and robots, which is a cornerstone of the human-centric approach of Industry 5.0. Human-robot collaboration leads to improved safety, efficiency, and productivity in manufacturing [19]. The integration of collaborative robots (cobots) and cybersecurity within Industry 5.0 emphasizes the importance of robust cybersecurity measures to protect the advanced interconnected systems used in Industry 5.0. The challenges and solutions related to ensuring the secure operation of cobots and other automated systems, crucial for maintaining trust and reliability in human-robot collaborations need to be addressed [20]. The advantages of cobots, such as flexibility, safety, and ease of integration, make them more suitable for human-centric production environments. The role of cobots in enhancing productivity while ensuring worker safety and comfort is explored, aligning with Industry 5.0's goals of creating more adaptable and human-friendly industrial processes [21].

Various emerging technologies that enhance human-machine interaction are pivotal for the paradigm shift to Industry 5.0. It discusses human-centered approaches, including virtual reality training and customer sentiment analysis, which are essential for creating more engaging and efficient work environments. The importance of these technologies in promoting a more interactive and responsive industrial ecosystem is explored [22]. The use of mixed-reality technologies to enhance interactions between humans and collaborative robots (cobots) provides a futuristic outlook on how mixed reality can be utilized to improve training, operational efficiency, and safety in Industry 5.0 environments. The integration of mixed reality supports the human-centric focus of Industry 5.0 by making technology more intuitive and accessible [23]. Magnetic levitation technology in industrial material handling systems provides a futuristic perspective on how this technology can improve efficiency, reduce friction and wear, and enable more precise handling of materials. The adoption of such advanced material handling systems is aligned with the goals of Industry 5.0 to create more efficient and sustainable industrial operations [24]. The idea behind the IIoT, or Industrial Internet of Things is the networked nature of different equipment, sensors, and devices that immediately gather, process, and distribute data. Integration of AI technologies with IIoT has been crucial to the advancement of this concept. AI provides intelligence, autonomy, and predictive capabilities to these interconnected systems. The incorporation of AI in IIoT has helped to process enormous volumes of data produced by connected devices into actionable insights.

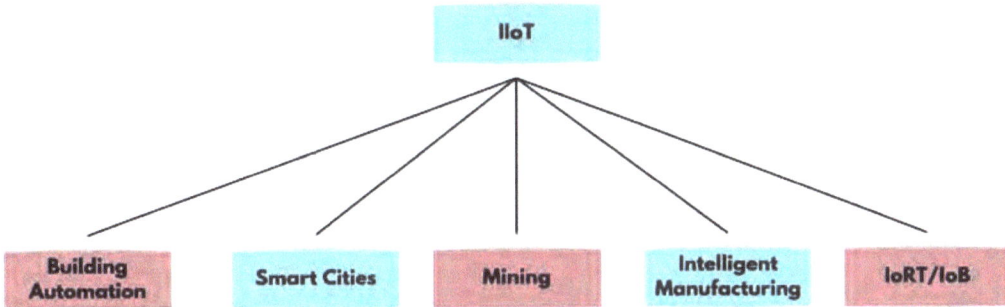

Fig. (4). IIoT application scenarios and data collection.

Artificial Intelligence has the capability to evaluate intricate data sets, detect trends, and derive valuable insights to enhance several facets of industrial processes. One of the most important uses of AI in the IIoT is predictive maintenance, where algorithms are used to continually track equipment performance, identify anomalies, and anticipate maintenance needs before breakdowns occur. This proactive strategy prolongs the life of vital assets while minimizing downtime and lowering repair costs. AI-powered IIoT systems enable autonomous operations, empowering machines to make intelligent decisions and real-time adaptation to shifting circumstances. By integrating AI algorithms with control systems, industrial processes can be optimized dynamically, adjusting parameters to maximize efficiency while ensuring product quality and safety standards are consistently met. Additionally, AI enhances quality control in manufacturing by employing computer vision and pattern recognition techniques to inspect products for defects with unparalleled accuracy and speed.

Furthermore, supply chain management is being revolutionized by AI-driven IIoT technologies that offer current insight into the state of inventory, demand projections, and logistical operations. By analyzing data from interconnected sensors and RFID tags, AI algorithms optimize supply chain processes, mitigate disruptions, and enhance operational resilience. Additionally, AI enhances energy efficiency in industrial settings by optimizing resource consumption, identifying energy-intensive processes, and recommending efficiency improvements to minimize environmental impact. Integration of AI with IIoT technologies represents a significant shift in industrial operations, unlocking unprecedented levels of automation, intelligence, and optimization across various sectors. By harnessing the power of AI-driven analytics, industrial enterprises can navigate the complexities of the digital era with agility, resilience, and competitiveness. The advantages of IIoT datasets are numerous, including facilitating data-driven decision-making, streamlining industrial processes, identifying irregularities, predicting faults, optimizing maintenance schedules, reducing energy usage, improving product quality, and enhancing customer satisfaction. With the incorporation of IoT datasets, industrial enterprises can boost efficiency, cut costs, and maintain a competitive edge.

CLOUD COMPUTING

The Internet-based offering of computer and networking solutions is known as cloud-based software. Because the server that is the archive, the database, social networking, analysis, and intelligence services are provided *via* the internet (the "cloud"), they are very flexible and priced. There are three main types of cloud-based computing services, namely Infrastructure as a Service (IaaS), Platform as a Service (PaaS), and Software as a Service (SaaS). Three primary models for cloud deployment are hybrid clouds, private clouds, and public clouds. The concept of "the design of anywhere and produce anywhere," could be accomplished in the early 2000s with the help of internet computing. Cloud manufacturing services are made possible by contemporary electronic and computing technologies, including sensors, GPS, IoT, RFID, cyber-physical networks, and cloud computing. Five major categories can be used to classify cloud manufacturing [25]. These include studies on resource definition and capabilities, architectural and platform design, service composition and selection, resource allocation and timetable, and service search and matching [26]. Cloud manufacturing is distinguished by its superior quality, affordability, dependability, and capacity for on-demand production. Moreover, additive manufacturing and manufacturing grids are two other modern manufacturing models that are being driven by cloud manufacturing. Data about the functioning state of manufacturing processes can be collected and analyzed in the cloud using IoT sensors. The upcoming generation of cloud manufacturing software, I5.0, is predicted to cater to more intricate and diverse needs in the fields of engineering, production, and organization. I5.0 is anticipated to handle a broader range of complex demands related to production, organization, and engineering in the future of cloud manufacturing systems. Further developments in 5G-based communications networks, EC characteristics, and AI/ML have unfolded new paths for significantly enhancing cloud manufacturing systems' future potential. An instrument or device known as a management data system, or MIS, facilitates the measurement of an organization's development by making it simpler to discern between the overall performance and the established objectives. Since MIS implementation involves the setup and upkeep of IT infrastructure, it is a costly process. This structure, like any well-managed information system, keeps things under control, enables a business to stop at any point if it gets stopped, and restarts when circumstances permit without erasing any work or duties completed in the prior stage. At each stage of MIS, we carried out each of the six processes in the suggested methodology as mentioned in Fig. (**5**).

Fig. (5). MIS procedures in cloud computing.

- **INSPECT:** This stage provides an organization with knowledge regarding what it needs and when, so explore it to make the greatest selection possible. Work requires to be done to secure internal permission and select the best provider.
- **GATHER:** This phase comprises obtaining the necessary information from the right source to fulfill the primary objective of the target audience.
- **TRANSFORM:** Adjustment to fulfill the requirements of the company

additionally enables users to fully utilize the data, this stage combines the data and turns it into a rich source of information.

- **CLEAN:** Tidying this stage involves the new cloud provider being managed as effectively as possible. It is necessary to handle the cloud provider and specifically clean up the vendor relationship because the firm will need to choose the new arrangement, particularly at the IT leadership level.
- **SECURE:** We employ a new replacement block cipher technique for security reasons. The algorithm makes use of a matrix's key, which produces a sequence when multiplied by a ternary vector and the product of a sign function [27]. The practical application of cloud-IoT technologies effectively addresses the obstacles encountered in the quest for global development. Cloud IoT has the potential to improve many facets of societal growth in addition to the financial and operational aspects of industry operations. The sensors linked to cloud-IoT systems could show improved levels of human well-being in several significant areas, including the healthcare industry, water supply network, agriculture industry, management of natural resources, and climate change.

In addition, S5.0 encourages I4.0 to address many personal and societal annoyances, thereby improving the quality of human life. However, because it necessitates constant effort, the hybridization of innovative technology with society will be a significant and challenging issue. It makes sense that in the not-too-distant future, many innovative technologies robots, AI, vast amounts of data, autonomous cars, and drone deliveries will be widely employed for the good of people and society as a whole [28].

OPPORTUNITIES IN I5.0

Demand forecasting is where AI in the management of inventory is expected to make revolutionary strides in the future. Even more precise forecasts may result from utilizing more advanced AI algorithms. AI has enormous potential for methodically analyzing large, complicated data sets to estimate demand [29]. This might result in more complex and dynamic inventory plans. AI in the management of inventory will help improve customer experience and personalization in the future [30]. This strategy could be applied to several industries, increasing client loyalty and happiness. Furthermore, I5.0 projects currently underway ensure that artificial intelligence (AI) optimizes industrial processes, producing speedier and more tailored products. International standards and rules are still being developed to codify I5.0, even though it is being embraced. Using AI and cognitive technologies, hyper-customization of all industrial processes will guarantee unique production solutions for every customer. When 5G, the Internet of Things (IoT), and AI are included, there are many exciting possibilities. The Inventory Management Potential is used for IoT and 5G-Enhanced Multi-Agent Systems [31]. With its real-time insights and increased efficiency, this connection has the potential to completely transform inventory tracking and management.

Future developments in IIoT will encompass a variety of cloud providers, all of which will be monitored remotely while keeping costs in mind [32]. The future is predicted to be filled with a significant amount of unrest by the writers, with IoT anticipated to have a bigger impact, provided users can lessen the constraints and consequences of the elements that influence IIoT adoption, as this review [33] mentions. Virtualizations of power plants are also made possible by IoT innovations. The expanding use of IoT could allow for the charging of electric cars to become even more widespread.

IoT has the capacity to expand and emerge as the next major development in equipment technology. In the active fast-paced climate action, technologies are being developed to make IoT setup simple. Creating "IoT as a Service" innovations is one tactic that will when carefully considered, broaden the scope of IoT applications [34]. Because cloud computing provides scalable and adaptable infrastructure resources, I5.0 companies can respond swiftly to shifting customer needs. Because of this flexibility, businesses can scale up or down operations as necessary, which increases productivity and lowers costs. Strong instruments for data analysis and machine learning (AI) applications are offered by cloud platforms. In I5.0, businesses may leverage AI and cloud analytics technologies to make data-driven decisions, optimize workflows, and glean valuable insights from vast datasets.

I5.0 stakeholders may communicate and collaborate more easily thanks to cloud-based collaboration technologies. These solutions increase productivity and efficiency by enabling remote employment, real-time collaboration, and information access from any location. IoT device and data stream management is made possible by cloud computing. In I5.0, instantaneous information from physical structures and environments may be gathered by IoT devices. This data can then be analysed and processed in the cloud to provide insights and improve operations. Initiatives aimed at implementing digital transformation in the manufacturing sector and S5.0 are facilitated by digital cloud computing. Traditional on-premises systems can be moved to the cloud to update IT infrastructure, increase agility, and spur innovation in enterprises.

CHALLENGES IN IMPLEMENTING I5.0 AND S5.0

However there is always the concern that traditional cloud designs will not be able to handle the massive volumes of data created, used by gadgets, and loaded during computation; keeping time constraints is another issue with current cloud designs

[35]. Additional difficulties result from the need to use data wisely after collection to obtain insights nevertheless, machine learning models and neural network algorithms can undertake automatic decisions, which is not necessarily reliable [36]. Security represents one of the possible issues. Security holes in managing heterogeneous data and employing cloud services for a range of consumer and commercial data management need to be double-checked as computing becomes more digital [37]. Scaling up user and manufacturing processes presents challenges that must be taken into consideration for personalized customer care that integrates human-robot collaboration. Furthermore, it is crucial to consider the ethical issues surrounding the application of AI to avoid any unfavorable consequences and repercussions on society. The use of AI by I5.0 and the resulting automation will open new channels for threat activities. Trusted execution is crucial for security in AI and ML tasks [38].

With the issues concerning Black Box AI, or the absence of openness associated with AI, it is important to consider the fourth industrial revolution when integrating AI with human intelligence [39]. There are several research obstacles preventing IoT solutions from being widely used in manufacturing:

- Data security and privacy concerns: many points of entry for a hacker are created when many connected devices are connected to the cloud.
- It may not always be possible to integrate legacy equipment due to the need for specialized patches.
- The need for open requirements: standards improving the interoperability of various systems, resulting in increased adoption and sustainable system functioning.
- Technological advances barriers: applications for manufacturing that require actual time feedback monitoring may not be able to manage the data transfer delay that may occur with IIoT. The expensive cost of sensor equipment and data ownership legal concerns are further obstacles.

Given the circumstances, new IoT applications for manufacturing present fresh research difficulties that call for technological solutions and standards that can capture massive real-time data streams and optimize processes throughout the manufacturing life cycle. The goal is to identify the ideal inventory level, reduce operating expenses, and consistently maintain that level [40]. Requirements must be provided in a timely way to ensure that production processes run smoothly. To track and manage inventory movements, resolve difficult successive decisions based on the learning with accessible data, and enable flexible learning under evolving conditions with no need for a preset environment model, most inventory control systems have used reinforced learning algorithms. Inventory control is one of the topics in inventory problems that has received the most research. The use of

AI is posing increasingly ethical and privacy-related issues. When using AI for inventory management, enormous amounts of data must be analyzed, some of which might include sensitive information [38]. The expense of AI solutions may prevent a firm from implementing them, especially for medium-sized businesses. Putting AI technology into practice can be expensive, requiring a lot of software, hardware, and human resources [41].

CONCLUSION

A significant step toward incorporating artificial intelligence (AI) into our socioeconomic environment is represented by I5.0 and S5.0. With AI-powered automated processes, scheduled upkeep, and personalized production revolutionizing traditional industries, I5.0 places a high priority on fostering cooperation between both humans and machines for increased productivity, creativity, and sustainability [42]. In the meantime, S5.0 aims to establish a society that is focused on people. AI technologies promote equitable sharing of technological advancements, improving healthcare, education, transportation, and governance to enhance inclusivity and mitigate social disparities. However, reaching I5.0 and S5.0's full potential requires addressing ethical, regulatory, and privacy concerns to ensure AI technologies serve the common good and protect fundamental human rights [43]. The convergence of these developments offers opportunities and challenges to enterprises, governments, and individuals in equal measure. By using AI's transformational potential responsibly, we can create a future where innovation thrives, communities flourish, and human well-being is prioritized. It is critical to approach this technological era with foresight, empathy, and a commitment to shaping a more inclusive and sustainable future for all.

REFERENCES

[1]   M. Ghobakhloo, H.A. Mahdiraji, M. Iranmanesh, and V. Jafari-Sadeghi, "From industry 4.0 digital manufacturing to industry 5.0 digital society: A roadmap toward human-centric, sustainable, and resilient production", *Inf. Syst. Front.,* no. 2, 2024.
      [http://dx.doi.org/10.1007/s10796-024-10476-z]

[2]   S. Mao, B. Wang, Y. Tang, and F. Qian, "Opportunities and challenges of artificial intelligence for green manufacturing in the process industry", *Engineering (Beijing),* vol. 5, no. 6, pp. 995-1002, 2019.
      [http://dx.doi.org/10.1016/j.eng.2019.08.013]

[3]   D. Mourtzis, "Towards the 5th industrial revolution: A literature review and a framework for process optimization based on big data analytics and semantics", *J. Mach. Eng.,* no. 9, 2021.
      [http://dx.doi.org/10.36897/jme/141834]

[4]   R.A. Cevallos, and C. Merino Moreno, "National policy councils for science, technology, and innovation: A scheme for structural definition and implementation", *Sci. Public Policy,* no. 12, 2020.
      [http://dx.doi.org/10.1093/scipol/scaa052]

[5]   C. Narvaez Rojas, G.A. Alomia Peñafiel, D.F. Loaiza Buitrago, and C.A. Tavera Romero, "Society 5.0: A Japanese Concept for a Superintelligent Society", *Sustainability (Basel),* vol. 13, no. 12, p. 6567, 2021.
      [http://dx.doi.org/10.3390/su13126567]

[6]   S. Huang, B. Wang, X. Li, P. Zheng, D. Mourtzis, and L. Wang, "Industry 5.0 and Society 5.0—Comparison, complementation and co-evolution", *J. Manuf. Syst.,* vol. 64, pp. 424-428, 2022. [http://dx.doi.org/10.1016/j.jmsy.2022.07.010]

[7]   F-Y. Wang, J. Yang, X. Wang, J. Li, and Q-L. Han, "Chat with ChatGPT on industry 5.0: Learning and decision-making for intelligent industries", *IEEE/CAA J. Auto. Sin.,* vol. 10, no. 4, pp. 831-834, 2023. [http://dx.doi.org/10.1109/JAS.2023.123552]

[8]   J. Leng, W. Sha, B. Wang, P. Zheng, C. Zhuang, Q. Liu, T. Wuest, D. Mourtzis, and L. Wang, "Industry 5.0: Prospect and retrospect", *J. Manuf. Syst.,* vol. 65, pp. 279-295, 2022. [http://dx.doi.org/10.1016/j.jmsy.2022.09.017]

[9]   D. Paschek, C-T. Luminosu, and E. Ocakci, "Industry 5.0 Challenges and Perspectives for Manufacturing Systems in the Society 5.0", *Advances in Sustainability Science and Technology,* pp. 17-63, 2022. [http://dx.doi.org/10.1007/978-981-16-7365-8_2]

[10]   D. Wang, C-T. Lu, and Y. Fu, "Towards automated urban planning: When generative and chatgpt-like ai meets urban planning", *arXiv:2304.03892 [cs],* 2023.

[11]   N. Rane, "ChatGPT and Similar Generative Artificial Intelligence (AI) for Building and Construction Industry: Contribution, Opportunities and Challenges of Large Language Models for Industry 4.0, Industry 5.0, and Society 5.0", *Social Science Research Network,* 2023. Available from: https://papers.ssrn.com/sol3/papers.cfm?abstract_id=4603221 [http://dx.doi.org/10.2139/ssrn.4603221]

[12]   C.T. Yang, H.W. Chen, E.J. Chang, E. Kristiani, K.L.P. Nguyen, and J.S. Chang, "Current advances and future challenges of AIoT applications in particulate matters (PM) monitoring and control", *J. Hazard. Mater.,* vol. 419, p. 126442, 2021. [http://dx.doi.org/10.1016/j.jhazmat.2021.126442] [PMID: 34198222]

[13]   J. Ahmad, M. Awais, U. Rashid, C. Ngamcharussrivichai, S. Raza Naqvi, and I. Ali, "A systematic and critical review on effective utilization of artificial intelligence for bio-diesel production techniques", *Fuel,* vol. 338, p. 127379, 2023. [http://dx.doi.org/10.1016/j.fuel.2022.127379]

[14]   Ekaterina Dmitrieva, G. Thakur, Pranav Kumar Prabhakar, A. Prakash, A. Vyas, and Y. Lakshmi Prasanna, "Edge Computing and AI: Advancement in industry 5.0- an experimental assessment", *Bio Web of Conferences/BIO web of Conferences,* vol. 86, p. 01096, 2024. [http://dx.doi.org/10.1051/bioconf/20248601096]

[15]   P.K.R. Maddikunta, Q-V. Pham, P. B, N. Deepa, K. Dev, T.R. Gadekallu, R. Ruby, and M. Liyanage, "Industry 5.0: A survey on enabling technologies and potential applications", *J. Ind. Inf. Integr.,* vol. 26, no. 100257, p. 100257, 2022. [http://dx.doi.org/10.1016/j.jii.2021.100257]

[16]   R. Ziatdinov, M. S. Atteraya, and R. Nabiyev, "The fifth industrial revolution as a transformative step towards society 5.0," *Societies,* vol. 14, no. 2, p. 19, 2024, [http://dx.doi.org/10.3390/soc14020019]

[17]   R. Tallat, "Navigating industry 5.0: A survey of key enabling technologies, trends, challenges, and opportunities", *IEEE Commun. Surv. Tutor.,* no. Jan, pp. 1-1, 2023. [http://dx.doi.org/10.1109/COMST.2023.3329472]

[18]   I. Affia, and A. Aamer, "An internet of things-based smart warehouse infrastructure: design and application", *J. Sci. Techno. Pol. Manag.,* vol. 13, no. 1, pp. 90-109, 2022.

[19]   R. Raffik, R. Sathya, V Vaishali, S Balavedhaa, and Jyothi Lakshmi N, "Industry 5.0: Enhancing human-robot collaboration through collaborative robots – A Review," Jun. 2023. [http://dx.doi.org/10.1109/ICAECA56562.2023.10201120]

[20]   A. Khalid, P. Kirisci, Z.H. Khan, Z. Ghrairi, K. D. Thoben, and J. Pannek, "Security framework for industrial collaborative robotic cyber-physical systems", *Comp. in Indus.*, vol. 97, pp. 132-145, 2018.

[21]   V. Yevsieiev and D. Gurin, *Comparative analysis of the characteristics of mobile robots and collaboration robots within INDUSTRY 5.0*, doctoral dissertation, European Scientific Platform, 2023,

[22]   M. Dhanda, B.A. Rogers, S. Hall, E. Dekoninck, and V. Dhokia, "Reviewing human-robot collaboration in manufacturing: Opportunities and challenges in the context of industry 5.0", *Robot. Comp. Integ. Manuf.*, vol. 93, p. 102937, 2025.

[23]   Z. E. Ahmed, A. H. Hashim, R. A. Saeed, and M. M. Saeed, "Monitoring of wildlife using unmanned aerial vehicle (UAV) with machine learning", *Applications of Machine Learning in UAV Networks*, IGI Global Scientific Publishing, 2024.

[24]   N. J. John, V. R. Ramesh, G. M. Arun, and R. R. Raj, "Magnetic levitation based industrial material handling systems: A futuristic review," *2023 2nd International Conference on Advancements in Electrical, Electronics, Communication, Computing and Automation (ICAECA)*, Coimbatore, India, pp. 1-6, 2023.
[http://dx.doi.org/10.1109/ICAECA56562.2023.10199189]

[25]   E.J. Ghomi, A.M. Rahmani, and N.N. Qader, "Cloud manufacturing: challenges, recent advances, open research issues, and future trends", *Int. J. Adv. Manuf. Technol.*, vol. 102, no. 9-12, pp. 3613-3639, 2019.
[http://dx.doi.org/10.1007/s00170-019-03398-7]

[26]   A. Raja Santhi, and P. Muthuswamy, "Industry 5.0 or industry 4.0S? Introduction to industry 4.0 and a peek into the prospective industry 5.0 technologies", *Intern. J. Inter. Desi. Manufac. (IJIDeM)*, vol. 17, no. 2, pp. 947-979, 2023. [IJIDeM]..
[http://dx.doi.org/10.1007/s12008-023-01217-8]

[27]   N. Kaushal Kishor, *Saxena, and D. Pandey, Cloud-based Intelligent Informative Engineering for Society 5.0*. Informa, 2023.
[http://dx.doi.org/10.1201/9781003213895]

[28]   K. Nath and C. Pandey, *Cloud-IoT Technologies in Society 5.0*, 2023.
[http://dx.doi.org/10.1007/978-3-031-28711-4]

[29]   Ö. Albayrak Ünal, B. Erkayman, and B. Usanmaz, "Applications of artificial intelligence in inventory management: A systematic review of the literature", *Arch. Comput. Methods Eng.*, vol. 30, no. 4, 2023.
[http://dx.doi.org/10.1007/s11831-022-09879-5]

[30]   M. K. Hossain Mondal, A. Mondal, S. Chakraborty, K. Shubhranshu, A. K. Jha, and M. K. Roy, "Advanced deep learning and nlp for enhanced food delivery: future insights on demand prediction, route optimization, personalization, and customer support," *2023 IEEE Renewable Energy and Sustainable E-Mobility Conference (RESEM)*, pp. 1-5, 2023.
[http://dx.doi.org/10.1109/RESEM57584.2023.10236091]

[31]   A.A. Mirani, G. Velasco-Hernandez, A. Awasthi, and J. Walsh, "Key challenges and emerging technologies in industrial iot architectures: A review", *Sensors (Basel)*, vol. 22, no. 15, p. 5836, 2022.
[http://dx.doi.org/10.3390/s22155836] [PMID: 35957403]

[32]   Y. Xu, J. Ren, G. Wang, C. Zhang, J. Yang, and Y. Zhang, "A blockchain-based nonrepudiation network computing service scheme for industrial IoT", *IEEE Trans. Industr. Inform.*, vol. 15, no. 6, pp. 3632-3641, 2019.
[http://dx.doi.org/10.1109/TII.2019.2897133]

[33]   B. Yu, and C. Xie, "Method for detecting industrial defects in intelligent manufacturing using deep learning", *Computers, Materials & Continua*, vol. 78, no. 1, 2024.

[34]   C. Yang, W. Shen, and X. Wang, "Applications of internet of things in manufacturing", *20th International Conference on Computer Supported Cooperative Work in Design.*, CSCWD: Nanchang,

China, pp. 670-675, 2016. Available from: https://doi.org/CSCWD.2016.7566069

[35]   Z. Fatima, M.H. Tanveer, Waseemullah, S. Zardari, L.F. Naz, H. Khadim, N. Ahmed, and M. Tahir, "Production plant and warehouse automation with IoT and industry 5.0", *Appl. Sci. (Basel),* vol. 12, no. 4, p. 2053, 2022.
[http://dx.doi.org/10.3390/app12042053]

[36]   R. Tallat, A. Hawbani, X. Wang, A. Al-Dubai, L. Zhao, Z. Liu, and S. H. Alsamhi, "Navigating industry 5.0: A survey of key enabling technologies, trends, challenges, and opportunities", *IEEE Commun. Surv. Tutor.,* vol. 26, no. 2, pp. 1080-1126, 2023.

[37]   N.V. Korneev, "Intelligent complex security management system FEC for the industry 5.0", *IOP Conf. Series Mater. Sci. Eng.,* vol. 950, no. 1, pp. 012016-012016, 2020.
[http://dx.doi.org/10.1088/1757-899X/950/1/012016]

[38]   R. Guidotti, A. Monreale, S. Ruggieri, F. Turini, F. Giannotti, and D. Pedreschi, "A survey of methods for explaining black box models", *ACM Comput. Surv.,* vol. 51, no. 5, pp. 1-42, 2019.
[http://dx.doi.org/10.1145/3236009]

[39]   H. Meisheri, N.N. Sultana, M. Baranwal, V. Baniwal, S. Nath, S. Verma, B. Ravindran, and H. Khadilkar, "Scalable multi-product inventory control with lead time constraints using reinforcement learning", *Neural Comput. Appl.,* vol. 34, no. 3, pp. 1735-1757, 2022.
[http://dx.doi.org/10.1007/s00521-021-06129-w]

[40]   Y. Chen, and M.I. Biswas, "Turning crisis into opportunities: how a firm can enrich its business operations using artificial intelligence and big data during COVID-19", *Sustainability (Basel),* vol. 13, no. 22, p. 12656, 2021.
[http://dx.doi.org/10.3390/su132212656]

[41]   S. Shakya, A. Liret, and G. Owusu, "Leveraging AI for asset and inventory optimisation," pp. 39–60, Aug. 2022,
[http://dx.doi.org/10.18573/book8.c]

[42]   J. Barata, and I. Kayser, "Industry 5.0 – past, present, and near future", *Procedia Comput. Sci.,* vol. 219, pp. 778-788, 2023.
[http://dx.doi.org/10.1016/j.procs.2023.01.351]

[43]   R.Y. Zhong, X. Xu, E. Klotz, and S.T. Newman, "Intelligent manufacturing in the context of industry 4.0: A review", *Engineering (Beijing),* vol. 3, no. 5, pp. 616-630, 2017.
[http://dx.doi.org/10.1016/J.ENG.2017.05.015]

CHAPTER 2

Industry 5.0 for Society 5.0: Roadmap Ahead

Pranjal S. Pandit[1,*], Grishma Y. Bobhate[2], Aniket S. Ingavale[3] and Saurabh Sathe[4]

[1] *Department of Computer Science & Engineering- Artificial Intelligence, Vishwakarma Institute of Technology, Pune, India*

[2] *Department of Computer Science & Engineering-Artificial Intelligence & Machine Learning, Vishwakarma Institute of Technology, Pune, India*

[3] *School of Engineering and Technology, DES Pune University, Pune, India*

[4] *Department of Computer Science, San Jose State University, San Jose, California, USA*

Abstract: Industry 5.0 is developing suggestive paradigms concerning the future development of societal structures and industrial operations concerning Society 5.0. Industry 5.0 is a complementary digital uprising of Industry 4.0, marked by a profound emphasis on human-machine collaboration, where human expertise and leading-edge technologies—particularly collaborative robots, or cobots —combine to formulate manufacturing systems that are both versatile and configurable to the competencies and creativity of human workers. In this phase, the main visible distinctive features are decentralized decision-making, the extensive use of augmented reality, and a commitment to sustainability.

Society 5.0, on the other hand, aims to create a civilization with a human-centric worldview. It is an association in which the actual world and the virtual space are blended, and data gives visibility between the two. It is about IoT, AI, and big data being put in the service of society to address major societal challenges and ensure they are growing basic lifestyle of people through inclusive and sustainable growth. The paradigm looks at smart cities, personalized healthcare, re-imagined education, and inclusive financial systems as ways technology can be used to address fundamental human needs—for quality of life, environmental stewardship, and equitable access to opportunities.

Keywords: Human-centric AI, Internet of things, Industry 5.0, Society 5.0.

INTRODUCTION

"Industry 5.0" and "Society 5.0" are terms coined to represent stages in the development of industries and societies. These ideas expand on the progress and

* **Corresponding author Pranjal S. Pandit:** Department of Computer Science & Engineering- Artificial Intelligence, Vishwakarma Institute of Technology, Pune, India; E-mail: pranjal.pandit@viit.ac.in

changes seen in societal models [1]. It is an addition to Industry 4.0, which deals with modernized technologies like computer vision, extensive statistical analysis, and machine learning algorithms [1]. Industry 5.0 signifies the new trends in technological development with a human-centric approach to regulate the new revolution in emerging fields. Recognizing essential elements of the human workforce helps characterize the potential benefits of marketing needs. This enhances productivity and employability [1]. Industry 5.0 focuses on economic value towards social development and encompasses innovation with new strategies.

In the evolution of Industry 5.0 and Society 5.0, it becomes evident that finding harmony between technological progress and human principles is crucial. These ideas shape discussions around equitable growth marking a shift towards an era where innovation goes beyond efficiency and advancement to deeply enhancing the human journey [2]. Thus, cognitive learning environments emphasize innovative business and the use of fully integrated services.

INDUSTRY 5.0

In the realm of Industry 5.0, we witness an advancement in practices. This stage expands on the framework established by its forerunners, such as Industry 1.0's focus on mechanization and Industry 4.0's introduction of manufacturing and digital integration. Industry 5.0 is unique in that it places a strong focus on encouraging human-machine cooperation [2]. Fig. (**1**) shown below represents the characteristics of Industry 5.0, which defines the workflow and encompasses the automation, integration, digitalization, customization, and the use of emerging tools.

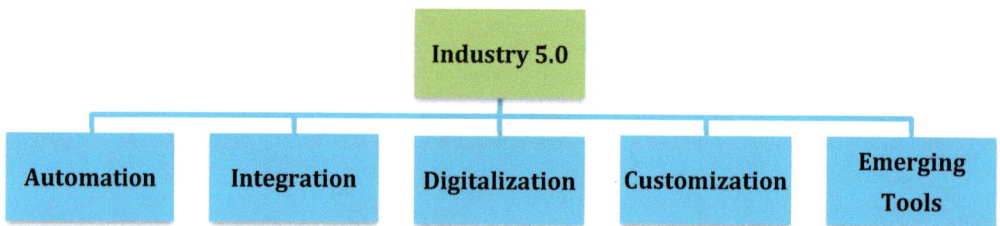

Fig. (1). Industry 5.0.

Industry 5.0 disintegrates the separation of employees from machines advocating for a harmonious blend of human creativity, skills, and intuition, with cutting-edge technologies [3]. Emerging tools like collaborative robots, known as cobots, take center stage in this era by working alongside humans in a beneficial partnership. The ultimate objective is to automate and develop manufacturing processes that not only excel in efficiency and technological sophistication but also prioritize the unique strengths of human laborers [3].

SOCIETY 5.0

"Society 5.0" depicts an outlook for human societies in the future, building on the digital revolutions that started with our agrarian and hunter-gatherer, ancestors and continued through the widespread adoption of digital tech and connectivity in Society 4.0 [4]. Society 5.0 is all about using technology to solve problems and make life better for everyone. It is like a super-duper version of our world, where everything is connected and super smart [4].

Statistical data analytics, computational intelligence, and blockchain technology will all play a big part in this [4]. Imagine a world where your fridge can tell your doctor that you're running low on milk, and your self-driving car can warn the traffic light that it needs a little greener time. It's like living in a giant, awesome, high-tech family that looks out for each other [4]. Fig. (**2**) shown below defines the adaptation of the newest technology to enhance productivity from peer-to-peer interaction and maintain the sustainability of the environment.

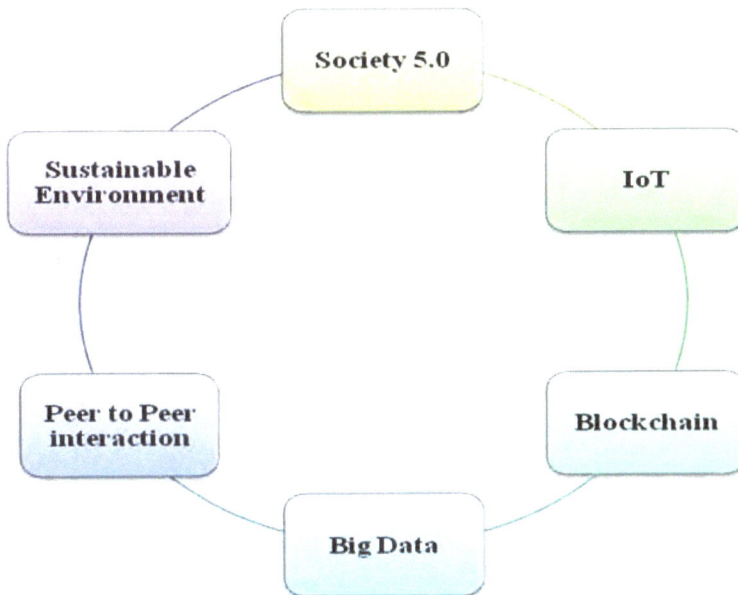

Fig. (2). Society 5.0.

Society 5.0 is used to incorporate appropriate software and hardware requirements with the help of the Internet of Things. Blockchain processes the data modality, and big data contains all information of users and maintains updated records to help in future needs. The important aspect of Society 5.0 is to restructure people's well-being, safeguard the surroundings, and make sure everyone has a fair shot at success. It's not just about having cool gadgets; it's about using technology to make our world a better place for everyone [5]. Industry 5.0 is all about this too, but for businesses and industries. It is about making sure that the way we do things doesn't just focus on making money, but also on being responsible and caring for people and the planet [5].

In a nutshell, Society 5.0 and Industry 5.0 are all about using technology to elevate people's standard of life while still keeping human values and needs at the forefront. They help us think about the future in a way that's smart, responsible, and inclusive.

EVOLUTION AND CHARACTERISTICS

The increasing demands in all sectors bring about a change in living mechanisms. The digital revolution has an impact on societal needs and economic growth. These characteristics have a significant effect on the structural usage of emerging technology in various sectors.

Evolution of Industry 5.0

Industry version 5.0 is still an experimental sector. It is widely regarded as the subsequent stage in the development of industrial processes. Here's a brief overview of the evolution leading up to Industry 5.0 [6] given in Fig. (3). The evolution of industry has taken place in every version to create new strategic developments for the betterment of society.

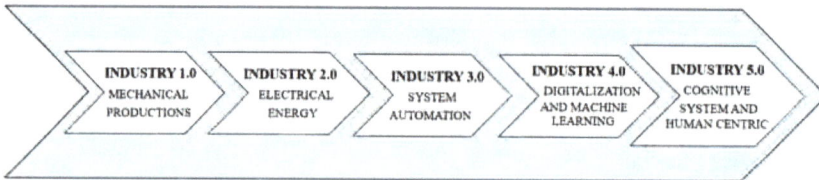

Fig. (3). Overview of industry evolution.

Industry 1.0: It is observed to be the 1st Industrial Revolution. Way back in the late 18th century, people started using water and machines to power machines. This was a big deal because it moved us away from manual craftsmanship and toward early industrial machinery. It started in England in the 18th century, when people started shifting from handicraft factories. So, this was seen as the mechanization of production and large-scale use of steam power. This revolution majorly impacted a few industries like glass, mining, agriculture, and the textile sector. Steam was known before, but when it first came into existence in this phase, it was considered to represent the most significant innovation ever. The usage of steam not only increased production but also human productivity on a large scale.

Industry 2.0: A few decades later, in the 19th century, we saw electric power and assembly lines. This allowed for mass production and standardization of products on a huge scale. It changed the game. This mainly happened in Germany, America, and Britain. This period is referred to as the "Technological Revolution" era. Here the steam power was replaced by electrical energy and oil. Electrical machinery was cheaper to work with, easier to maintain and had greater productivity than steam-powered equipment. The machines used in this era were cost-effective and needed less human effort. Mass production was considered a standard practice during this era. Labor requirements decreased, which led to the loss of jobs as their work was replaced by machines. This was accompanied by an extensive telegraph and railroad network, which allowed faster communication and information transfer.

Industry 3.0: In the late 20th century, computers and automation tech took over. This is referred to as the "Digital Revolution" or the "first computer era." We started using electronic systems in our manufacturing processes, and stuff like programmable logic controllers (PLCs) made things even smarter. This was the beginning of digitalization. However, they still relied on human intervention and input. Simple yet large computers were developed. PLCs have a greater effect on software systems and can handle hardware devices. They stimulate the rise in market demands for software development during a certain period of time. The software systems made a lot of management processes possible, for example, activities like product tracking, inventory management, scheduling product flow, shipping logistics, enterprise resource planning, and so on. Other electronic machines that were developed in this era included IC chips, DLSs, and MOSFET semiconductors, together with the developments that they originated through, including microprocessors, smartphones, personal computers, and the internet.

Industry 4.0: Early in the 21st century, we saw the rise of smart manufacturing. This was all about combining physical and digital technologies, which took things up a notch. Cloud computing, big data, artificial intelligence, and the Internet of Things, also known as IoT, all are combined to build things faster, more efficiently, and more connected than ever before. This is currently being implemented in our modern world. Cyber-physical industrial platforms are extensively connected, and this is attributable to this fourth revolution. Owing to this concept, manufacturing is now entirely automated, and component and interpersonal interaction can be accomplished in smart manufacturing. The manufacturing sector has been entirely stored digitally as a result, which has simplified the flow of information to the right individuals at the appropriate times. The fourth revolution also involves a stronger emphasis on environmental and sustainability concerns. Sustainability is a chance to improve business and manufacturing procedures in addition to being an imperative to become more environmentally conscious while preserving natural resources for the next generation.

Industry 5.0: With this revolution, we moved toward human-centric automation. Relying upon the principles of Industry 4.0, this emphasizes human-machine interactions more than before. Technologies like collaborative robots (cobots) are used to make workers jobs easier and more enjoyable. The goal is to make sure that humans are still an important part of the process, even as technology advances.

Characteristics of Industry 5.0

Industry 5.0 is the newest current production process and industrial revolution approach that prioritizes human value, emerging technologies, and sustainable development. Here are some characteristics taken into account as follows:

Human-Centric Technique: Industry 5.0 is all about bringing people back into the tech loop. It focuses more on integrating humans with advanced technologies in manufacturing, unlike Industry 4.0, which was all about automation and data exchange [6]. In this approach, humans are kept at the center of production. Along with sustainable development, the overall well-being of the workers is prioritized. This is important to improve individual skills and competence concerning digital technologies.

Collaborative Robotics (Cobots): One of the important aspects of the sector of Industry 5.0 is using cobots, which are robots that work alongside humans to make factories safer and more efficient [6]. These machines have evolved as they are equipped with sensors, actuators, and AI-powered controllers that enable them to work next to humans without causing intrusion. Cobots are very versatile, easy to program, safe, and intuitive to use. Their use alongside humans can help unlock innovations.

Customization and Flexibility: This new industry wants factories to be more flexible and able to change according to what's needed. This is different from Industry 4.0, which was more about making things the same way all the time [7]. This feature enables a wide range of product delivery in smaller quantities so that customer satisfaction is met as a priority. This is me with high safety standards and improved sustainability.

Decentralized Decision-Making: Decentralised decision-making is made easier with technologies like edge computing, which lets factories respond faster to changes and makes it so they don't have to rely on one person to make all the calls [7]. This makes the system more transparent and verifiable and enables faster incorporation of demands and changes as per customers needs.

Sustainability: Industry 5.0 is about being environmentally friendly and not harming people or the planet. This means using resources wisely and not making stuff that's going to upset the environment or people [7]. The move to Industry 5.0 is all about finding a balance between technology and human skills. It is not just about machines doing everything, but about working together more smartly. It is not figured out yet, but we will keep learning and growing as technology gets better [7].

Characteristics of Society 5.0

Society 5.0 has a greater effect on the development of socioeconomic needs and sets the different aspects of a better standard of living. Some characteristics of Society 5.0 highlight the following points:

Human-Centric Society: Like Industry 5.0, Society 5.0 is all about using technology to help people. It is about finding new ways for technology to make our lives better and more comfortable [7]. Human-centered methodology focuses on users' needs and requirements to build smart systems enhanced by data-driven technology.

Integration of Physical, Real, and Virtual Worlds: In Society 5.0, augmented reality, cognitive intelligence, and the embedding of software requirements in hardware ultimately merge to strengthen accessibility and simplify life [8].

Sustainable Development: This new society wants to find solutions to problems like pollution and an aging population by using technology [9]. Sustainable development is a holistic approach to addressing the interconnected challenges of poverty, inequality, environmental degradation, and climate change to create a more prosperous, equitable, and resilient world for present and future generations.

Inclusive Growth: Society 5.0 aims to make sure that everyone benefits from new technologies, not just a few people. It wants to use technology to bridge the gap between different groups of people [9]. By addressing social issues and focusing on economic progress, social innovation creates a smart society with advanced analytic frameworks.

It is important to remember that the stream concepts of Industry 5.0 are still being talked about and might change over time to the standardization of Society 5.0. Also, it is not just about having the technology; people need to be willing to use it and accept it for these ideas to work. And, of course, government policies play a major role in this aspect [9].

ISSUES AND CHALLENGES

While Industry 5.0 with the uprising sector of Society 5.0 is supposed to proceed with new advancements and benefits, they also come with some challenges that need to be faced to make sure they work well. Some of the challenges that come with Industry 5.0 to recreate the structural analysis of Society 5.0 are as follows:

Challenges of Industry 5.0 for Society 5.0

Industry 5.0 for the betterment of society 5.0 has given better results with tools, methods, and concepts in business strategy and organizational development. Some factors can influence different transformations in the competitive world.

Human-Automation Integration: Making people to work together with robots can be hard because you have to make sure they're safe, train them well, and make sure they're okay with the new technology [10]. In this Industry 5.0 technology, humans and machines are cooperating to achieve greater liability and improvement in terms of scalability and effectiveness in development. Due to high digitization, Industry 5.0 increases employability and customization in product design. This enrolls for a human-centric approach and production level.

Skill Gaps: The move to Industry 5.0 might lead to people not having the right skills for their jobs. To fix this, we need to keep training and teaching workers new things so they can handle the new technology [10]. However, an adaptation of the newest technology and hands-on experience requires integrating human skills and support for future industrial growth.

Data Security and Privacy: As things get more connected and data is shared more, there is a bigger risk of cyberattacks. To keep sensitive information safe and protect people's privacy, we need to find better ways to keep data safe [10]. It is crucial to build trusted systems for protecting confidential information and ensure the adoption of necessary regulations. Therefore, Industry 5.0 demands cybersecurity measures to safeguard individuals' information and emphasizes industry data-driven decisions.

Standardization and Interoperability: Industry 5.0 is based on a lot of different technologies and systems, so we need to find ways for all of them to work well together and not conflict [11]. It involves automated machines, smart systems, and cognitive computing that need to follow standard rules and regulations for handling different issues and interoperability.

Infrastructure Investments: Moving to Industry 5.0 might need a lot of money to upgrade and modernize old industrial stuff. This could be hard for small businesses or businesses in places that are not as developed [11]. This offers sustainability to increase the production level on a wider scale and encourages startups or small businesses. This leads to implementing new approaches and finding different ways to reach out at the global level.

Challenges of Society 5.0

The latest framework of Society 5.0 embarks on a new set of challenges to producing a smart society where technology can resolve various social issues.

Ethical Concerns: As technology becomes a bigger part of our lives, there are questions about how to use it in a fair and safe way, like keeping people's data private and making sure we do not spy on them too much [12]. Society 5.0 is concerned with maintaining a balance between creativity and ethics. Ensuring fairness, transparency, and accountability in technological developments is essential to prevent widening inequalities.

Inequality and Access: It is important that everyone can take part in the benefits of Society 5.0, but sometimes not everyone has the same chances. This can lead to more inequality in society [12]. In Society 5.0, addressing inequality involves recognizing and mitigating disparities related to income, education, healthcare, employment, and other socio-economic factors. It encompasses issues such as wealth inequality, social mobility, and disparities in access to essential services. Access is termed as the utilization of resources, services, and opportunities by an individual or people. In order to meet advancement and a sustainable society, Society 5.0 provides access to essential services such as education, healthcare, transportation, information, and digital technologies.

Resistance to Change: Moving to Society 5.0 means bringing big changes in how people live and work. Some people might not want to try new things or change how they do things, which could slow down the change [12].

Environmental Impact: While Society 5.0 wants to be more sustainable, using more technology and having more electronic stuff can have bad effects on the environment, like more electronic waste and using too much energy [12]. Thus a strong foundation for an effortless transition diversified digital solutions with green technology. Leveraging technologies like remote sensing, satellite imagery, and data analytics enables real-time monitoring of environmental parameters such as air quality, water quality, and biodiversity and has an adverse impact on climate change.

Regulatory Frameworks: It is hard to make rules and guidelines that will help people use technology in the right way. It is important to work together with different groups to find the best ways to do this [13]. This ensures regulatory bodies may struggle to adapt regulations to new technologies, address emerging risks, and ensure compliance across different sectors and industries. By addressing these issues, commitment to ethical governance and principles navigates the complexities of regulatory frameworks.

To deal with these challenges, it is important for everyone to work together. Governments, businesses, schools, and regular people all have a part to play in making sure Industry 5.0 and Society 5.0 work well. As these ideas keep changing and growing, it is important to keep talking and working on ways to deal with the challenges that come with them [13].

Solutions to Industry 5.0

Industry 5.0 promotes greater flexibility and increased collaboration with a higher degree of automation, which thoroughly includes artificial intelligence and optimal integration of innovation and creativity. To address these challenges, several solutions need to be provided for effective communication in a human-centric approach, sustainable adaptation, and development in various sectors. Fig. (**4**) describes the various factors that can be helpful to provide effective solutions to Industry 5.0, such as human intelligence used for innovation and ideas, advanced integration algorithms applied in various areas, sustainability for maintaining economic growth, security concerned with data privacy and integrity, system recovery applied to resilience to loss, and regulatory compliance ethics to be followed by every organization and institution to keep confidential data records secure, accessible, and in standard format [14].

Human Intelligence

Human-centered technology focuses on human demands to increase productivity and creativity to give efficient business solutions. Humans can collaborate their work with machines to recognize what technology has to improvise such that it satisfies customer-centric operations [15]. With this, Industry 5.0 greatly engaged the deployment of next-generation technology for the customization of products as per the customer's needs. Human Intelligence plays an important role in utilizing different automation tools to bridge the gap between skilled labor and the development of highly manufactured systems.

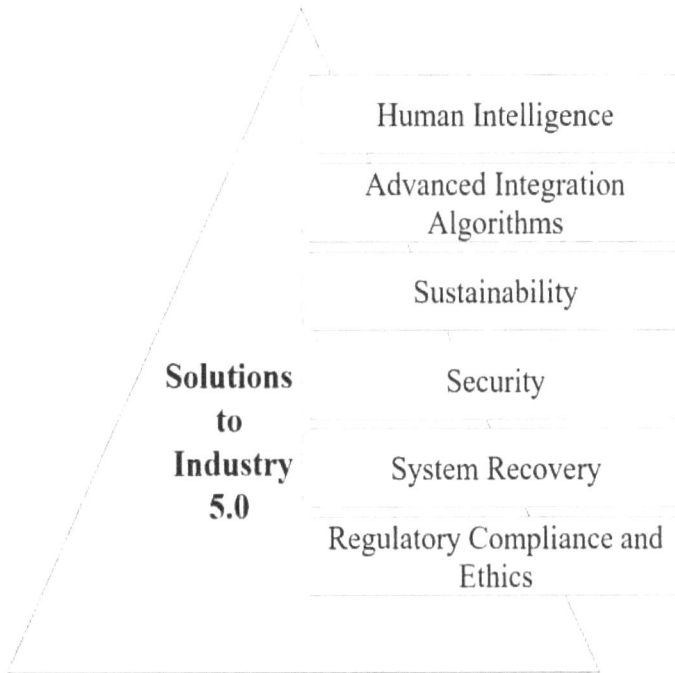

Fig. (4). Solutions to Industry 5.0.

Advanced Integration Algorithms

Industry 5.0 working module is able to integrate different high computing models used in a variety of domains like artificial intelligence, cloud computing, big data analytics, supply chain management, blockchain, and IoT [15]. Modernized and innovative techniques in machines encourage business organizations to enhance decision-making and predict results. Smart edge-cutting technologies imply automated processes and optimized production rates. This use of technology is beneficial to encourage the rise of subsequent industrialization in the community.

Industry 5.0 grows with the health sector field, which relates to analyzing and detecting the problems occurring in the body. This is taken into consideration for the development of creative technologies that personalize a reliable and portable device or applications to be handled by the patients. Integrated Artificial Intelligence algorithms have the capability to monitor and maintain health records.

Sustainability

Apart from technological advancement for society 5.0, sustainability provides effective solutions that run on renewable energy [15]. This may involve reducing carbon emissions and fossil fuels, analyzing the sources of raw materials and waste generation, and adopting new business models. This can help decrease environmental impact and lead to achieving objectives of increasing smart manufacturing practices and emphasizing economic principles.

Security

Industry 5.0 is an asset for the customization of products using new advanced techniques and tools, production digitalization, and expertise in the new era of Artificial Intelligence, as well as fostering a culture of innovation and collaboration [16]. To increase productivity and reduce complexity in data sharing between humans and machines, data integrity and confidentiality are important to ensure unauthorized access. Industry 5.0 must ensure the adoption of new security rules and principles to protect sensitive information.

System Recovery

Society 5.0 should have a key component of resilience in Industry 5.0. To provide economic growth and greater flexibility in the use of digital technologies and methodologies, resilience incorporates system functioning and proper usage of machines [17]. Thus, Industry 5.0 promises the contribution and advancement to ensure the stability of the workforce and understanding of more resilient industries for the development of Society 5.0.

Regulatory Compliance and Ethics

Industry 5.0 should prioritize and create specific regulations to avoid disruption in data privacy and security. This helps to apply ethics standards to balance innovation and promote the achievement of societal goals [17]. International collaboration also regulates ethical standards to establish work proficiency between education and industrial organizations.

APPLICATIONS

Industry 5.0 and Society 5.0 are ideas that concentrate on humans working with advanced technology to make things better. These ideas are still being worked on, but there are some possible ways they could be used:

Applications of Industry 5.0

Cobots: Industry 5.0 wants to use robots that can work with humans in factories. These cobots can help things run more smoothly and safely and make it easier for humans and machines to work together [17]. Cobots have the capability to collaborate with various manufacturing machines to understand the processing and deliver a good quality product.

Customized and Flexible Manufacturing: This idea is about making manufacturing processes more flexible and able to make different things, based on what people want. This could lead to more personalized products [17].

Human-Machine Integration: This means using technology like VR and AR to help people work with machines better. This can help people make decisions faster and solve problems [18]. It provides the autonomy for research and development in different sectors.

Decentralized Decision-Making: The goal of Industry 5.0 involves distributed decision-making so that decisions can be made quickly and without needing to go to a lot of people. This can be done with technology like edge computing, which lets computers process data right where it's collected.

Sustainable Manufacturing: This is about finding ways to use resources more wisely, make less waste, and find ways to make factories more environmentally friendly. This includes conservation of energy, recycling and reusing strategies, reduction of waste, and resource optimization, which are all vital elements of ecologically sustainable manufacturing.

Applications of Society 5.0

Smart Cities: Society 5.0 aims to establish urban areas where technology is used to make life better for people who live there. This could mean having better infrastructure, public transportation, and city planning [18]. This enables people to be highly involved with innovation and creation.

Healthcare Innovation: Society 5.0 could change healthcare by using technology like telemedicine, wearable devices, and computers to help doctors make diagnoses. This could lead to more personalized and efficient healthcare, with a focus on prevention and taking care of patients [18].

Education Transformation: Technology could change how we learn by giving us more personalized and interactive educational experiences, like online classes and tools that help us learn [18]. The educational sector modifies the learning ability, teaching skills, and methodologies to sympathize with the developing young minds.

Inclusive Financial Systems: Society 5.0 could create financial systems that are fairer and more accessible to everyone, using technology like blockchain and digital currencies [18]. This will severely process the inclusivity and accessibility to empower individuals to utilize online financial services effectively.

Technology could help us keep an eye on the environment and find ways to protect it. This could mean using sensors to detect pollution, finding ways to farm more sustainably, and managing natural resources better [18].

It's important to remember that these ideas are still being worked on and might change as technology gets better. There are also other things to consider, like what people think about these ideas and what laws and ethical rules need to be followed. As we learn more, we'll probably find even more ways to use technology to make our lives better.

CASE STUDY

Industry 5.0 aims to perform a cumulative effect on the industrial revolution in the production, manufacture, and different areas of the high-cognitive world [18]. The industrial and technical development of Industry 5.0 manages creative and innovative interfaces among humans and machines to give high-quality precision and general welfare to Society 5.0. In this case study, Industry 5.0 focuses more on societal challenges and the revolution of the education system.

Society 5.0 is the main approach to building collaborative, human-specific approaches to developing highly integrated algorithms and solving social issues effectively. The education system has a great interest in the development of society. Education 5.0 conforms with society 5.0 to provide students with great access to learning resources, and an understanding of personalized learning experiences with the integration of emerging technologies [19]. Due to the rapid increase in technological demand, the education system should promote new training programs, specialist courses, curricula, and emerging learning tool paradigms. This will make students thrive for digital competencies in the Industry 5.0 era. Educationists, teachers, and experts will understand significant investment in advanced technologies and facilitate a collaborative environment.

Fig. (5) represents a brief explanation of different elements that can be responsible for evaluating Education 5.0. One of the main challenges in the 5.0 industry in education is the adaptation of learning methodologies. This incorporates smart classrooms, a revolutionized pedagogical approach, and assessment methods. Every learner or student should have access to educational online resources, virtual assistants, and AI-based learning models to enhance their learning reflection. The continuous learning method helps us to update in this evolving industry [20]. The use of AI tools and machine learning algorithms adjusts the curriculum and difficulty levels of students' performance. Smart classrooms, AR, and VR programs simulate the understanding of the concepts with interactive displays and enhance the use of technology.

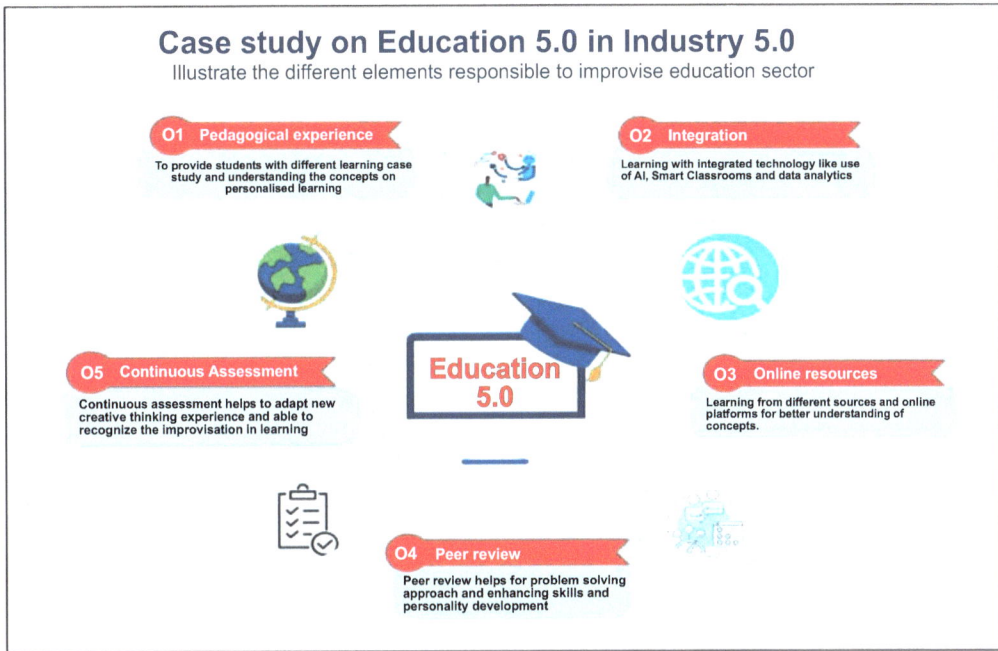

Fig. (5). Case study on Education 5.0 in Industry 5.0

The formation of an intelligent education system will gain advantages through Education 5.0 in terms of learning integrated platforms, skills development, and socioeconomic growth. Different effects of industrial 5.0 on education can leverage the redesigning and implementation of a smart intelligent society that characterizes the creative immersive learning contribution [20]. Thus, the transformation of the education system develops critical thinking, problem-solving skills, the ability to adapt to the newest technical skills, and cultural awareness. Education 5.0 withstands the student's perspective to face uncertainties and is enriched with a skill set. Therefore, the integration of Education 5.0 with Society 5.0 will expand personal life experience and form a better standard of living.

CONCLUSION

The implemented idea of the fourth industrial revolution is referred to as Industry 5.0, which utilizes automation techniques with the latest technology to simplify the learning process and focus on human-centric value. In this way, Society 5.0 employs novel technologies to keep an equilibrium between economic development and social values. Thus, the learning capability and integrated design tools analyze the efficient sustainable development and predict the performance strategic plan. Therefore, building human-computer interaction with innovative ideology motivates the intuitive understanding between humans and machines. The cultivation of learners' training competency abilities is an essential part of Education 5.0. Future studies on other applications can increase interest in studying and illustrate the real-time contributions to social, economic, and financial growth. Hence, India may emerge as an innovator in cognitive and interactive manufacturing systems by incorporating Industry 5.0 with these initiatives and endeavors.

REFERENCES

[1]   R. Raffik, and R.R. Sathya, "Industry 5.0: Enhancing human-robot collaboration through collaborative robots – a review", *2nd International Conference on Advancements in Electrical, Electronics, Communication, Computing and Automation (ICAECA),* p. 16, 2023.

[2]   S. Huang, B. Wang, X. Li, P. Zheng, D. Mourtzis, and L. Wang, "Industry 5.0 and society 5.0—Comparison, complementation and co-evolution", *J. Manuf. Syst.,* vol. 64, pp. 424-428, 2022.
[http://dx.doi.org/10.1016/j.jmsy.2022.07.010]

[3]   R.Y. Zhong, X. Xu, E. Klotz, and S.T. Newman, "Intelligent manufacturing in the context of industry 4.0: A review", *Engineering (Beijing),* vol. 3, no. 5, pp. 616-630, 2017.
[http://dx.doi.org/10.1016/J.ENG.2017.05.015]

[4]   Cabinet office, Council for science, technology and innovation, society 5.0, Council for science, technology and innovation, 2024.

[5]   C. Patrikakis, and K. Law, "Society 5.0: Human centric, decentralized, and hyperautomated", *IT Prof.,* vol. 24, no. 3, pp. 16-17, 2022.
[http://dx.doi.org/10.1109/MITP.2022.3177281]

[6]   M. Breque, L. De Nul, and Petridis, "Industry 5.0: Towards a sustainable, human-centric and resilient european industry," *European Commission, Directorate-General For Research and Innovation,* Luxembourg, 2021.

[7]   E. Flores, X. Xu, and Y. Lu, "Human Capital 4.0: A workforce competence typology for Industry 4.0", *J. Manuf. Tech. Manag.,* vol. 31, no. 4, pp. 687-703, 2020.
[http://dx.doi.org/10.1108/JMTM-08-2019-0309]

[8]   F. Longo, A. Padovano, and S. Umbrello, "Value-oriented and ethical technology engineering in industry 5.0: A human-centric perspective for the design of the factory of the future," *Appl. Sci.,* 10(12), 4182, pp 1-25, 2020.
[http://dx.doi.org/10.3390/app10124182]

[9]   S. Nahavandi , "Industry 5.0—A human-centric solution," *Sustainability,* 11(16), 4371, pp. 1-13, 2019.
[http://dx.doi.org/10.3390/su11164371]

[10]   Y. Lu, J.S. Adrados, S.S. Chand, and L. Wang, "Humans are not machines—anthropocentric human–machine symbiosis for ultra-flexible smart manufacturing", *Engineering (Beijing),* vol. 7, no. 6, pp. 734-737, 2021.
[http://dx.doi.org/10.1016/j.eng.2020.09.018]

[11]   G.Q. Huang, B. Vogel-Heuser, M. Zhou, and P. Dario, "Digital technologies and automation: The human and eco-centered foundations for the factory of the future [tc spotlight]", *IEEE Robot. Autom. Mag.,* vol. 28, no. 3, pp. 174-179, 2021.

[http://dx.doi.org/10.1109/MRA.2021.3095732]

[12]   D. Paschek, A. Mocan, and A. Draghici, "Industry 5.0—The expected impact of next industrial revolution", *Proceedings of the Makelearn and TIIM International Conference,* pp. 125-132, 2019.

[13]   L. Wang, "A futuristic perspective on human-centric assembly", *J. Manuf. Syst.,* vol. 62, pp. 199-201, 2022.
        [http://dx.doi.org/10.1016/j.jmsy.2021.11.001]

[14]   X. Xu, Y. Lu, B. Vogel-Heuser, and L. Wang, "Industry 4.0 and industry 5.0—inception, conception and perception", *J. Manuf. Syst.,* vol. 61, pp. 530-535, 2021.
        [http://dx.doi.org/10.1016/j.jmsy.2021.10.006]

[15]   V. Khullar, V. Sharma, M. Angurala, and N. Chhabra, *Artificial intelligence and society 5.0: Issues, opportunities, and challenges. chapman and hall/CRC press.* Taylor & Francis Group, 2024.

[16]   S. Joglekar and S. Kadam, "Industry 5.0: Analysis, applications and prognosis," The Online Journal of Distance Education and e-Learning, vol. 11, no. 1, pp 257-264, 2023. Available from: https://www.researchgate.net/publication/368653213_Industry_50_Analysis_Applications_and_Prognosis

[17]   A. Adel, "Future of industry 5.0 in society: human-centric solutions, challenges and prospective research areas", *J. Cloud Comput.,* vol. 11, no. 1, p. 40, 2022.
        [http://dx.doi.org/10.1186/s13677-022-00314-5] [PMID: 36101900]

[18]   R. Trivedi, "The role of industry 5.0 in education 5.0 in indian perspective", *Int. J. Innov. Res. Technol.,* vol. 10, no. 5, pp. 133-137, 2023.

[19]   M. Al-Emran, and M.A. Al-Sharafi, "Revolutionizing education with industry 5.0: Challenges and future research agendas", *Int. J. Inf. Technol. Lang. Stud.,* vol. 6, no. 3, pp. 1-5, 2022.

[20]   D.G. Broo, O. Kaynak, and S.M. Sait, "Rethinking engineering education at the age of industry 5.0", *J. Ind. Inf. Integr.,* vol. 25, 2022.

CHAPTER 3

Unleashing Opportunities: Bridging Research Gaps in High-Performance Computing for Holistic Decision Support in Clinical Informatics

Santosh Kumar[1,*] and **Tanmay Anil Rathi**[2]

[1] *Department of Artificial Intelligence & Data Science, Vishwakarma Institute of Technology, Pune, Maharashtra 411048, India*

[2] *IT Department, New York University, New York, NY 10012, USA*

Abstract: Objective: This systematic literature review navigates the landscape of high-performance computing (HPC) applications in clinical informatics, focusing on holistic decision support. The study aims to identify and address research gaps, optimize computational algorithms, and explore multi-modal data integration challenges.

Methods: Employing rigorous inclusion criteria, we systematically review existing literature to analyze the current state of HPC in clinical decision support. Methodological details encompass criteria for study selection, the search strategy employed, and the synthesis and analysis of data.

Findings: The review uncovers critical research gaps, notably in the scalability of computational algorithms and the integration of diverse healthcare data types. Optimization techniques and parallel computing approaches emerge as pivotal strategies to bridge these gaps. Challenges in multi-modal data integration and algorithmic approaches for comprehensive data analysis are explored.

Implications: Insights gleaned from real-world applications and case studies contribute to understanding successes and challenges in HPC implementation. Evaluation metrics for performance assessment are synthesized, providing a foundation for future research directions and emerging trends in HPC for clinical informatics.

Conclusion: This systematic review advances our comprehension of the research landscape, offering a roadmap for optimizing HPC applications in clinical decision support. The findings contribute to the ongoing discourse on leveraging computational power for holistic healthcare solutions.

* **Corresponding author Santosh Kumar:** Department of Artificial Intelligence & Data Science, Vishwakarma Institute of Technology, Pune, Maharashtra 411048, India; E-mails: dssant@gmail.com, santosh.kumar@viit.ac.in

Parikshit N. Mahalle, Gitanjali R. Shinde, Namrata N. Wasatkar & Prashant R. Anerao (Eds.)

Keywords: Clinical informatics, Computational algorithms, Decision support, High-performance computing (HPC), Multi-modal data integration, Optimization techniques.

INTRODUCTION

In the ever-evolving landscape of healthcare, the research paper titled "Unleashing Opportunities" takes a pioneering stance at the convergence of High-Performance Computing (HPC) [1] and holistic decision support within clinical informatics. This comprehensive exploration goes beyond computational methodologies to envision proactive strategies that address prevailing research gaps.

Structured meticulously, the paper navigates through foundational aspects of HPC's role in healthcare, proposing strategic optimization approaches [2] while projecting forward to anticipate emerging trends. The term "holistic" anchors the theme, emphasizing comprehensive decision support across diverse healthcare data types like electronic health records, genomic data, and wearable sensor information [3].

More than an academic exercise, "Unleashing Opportunities" is a call to action, inspiring future research endeavors and aiming to catalyze change in clinical decision-making practices. It extends beyond existing understandings, contributing to the evolution of clinical decision support. By engaging with the challenges and promises of HPC in healthcare informatics, this paper seeks to uncover transformative opportunities.

Through a proactive engagement with the challenges and promises of HPC in healthcare informatics, this paper seeks to uncover opportunities that can reshape the very fabric of decision-making in clinical settings.

Background and Context

In the dynamic landscape of healthcare informatics, the convergence of High-Performance Computing (HPC) and holistic decision support represents a pivotal frontier [4]. The escalating complexity and abundance of healthcare data, spanning electronic health records to genomic information, demand advanced computational methodologies [5].

This research unfolds within a healthcare ecosystem grappling with diverse data types and a growing need for sophisticated decision support systems.

The promise of transformative change in clinical decision-making accompanies the advent of HPC [6]. However, challenges and research gaps currently impede its full integration into healthcare [7]. The title, "Unleashing Opportunities," embodies the aspiration to not only identify these gaps but to actively bridge them, unlocking new avenues for comprehensive decision support.

This context is embedded in a field where precision, speed, and nuanced decision-making are paramount [8]. Electronic health records store extensive patient information, medical images provide visual insights, genomic data offers personalized perspectives, and wearable sensors generate real-time health data [9]. Navigating this complex landscape requires computational methodologies beyond conventional capabilities, positioning HPC as a transformative solution [10].

This research emerges at a critical juncture in healthcare, where the fusion of computational power with decision support has the potential to redefine patient outcomes and treatment strategies [3].

"Unleashing Opportunities" is not merely a research endeavor; it is a strategic initiative to propel healthcare informatics into a new era, where the synergy between HPC and decision support becomes a transformative force for improved patient care and outcomes.

Significance of High-Performance Computing in Clinical Informatics

In the evolving landscape of clinical informatics, the integration of High-Performance Computing (HPC) represents a transformative shift, profoundly impacting decision-support paradigms [3]. The research paper titled "Unleashing Opportunities" underscores HPC's profound significance in addressing critical research gaps and advancing comprehensive decision support in clinical informatics.

Swift Data Processing and Analysis: HPC's ability to swiftly process vast volumes of healthcare data is foundational in time-sensitive clinical settings [10]. Rapid processing and analysis enable more responsive interventions and timely decision-making.

Precision and Accuracy in Decision-Making: Precision is paramount in clinical decisions [11], and HPC ensures highly accurate analyses, reducing the margin for error [8]. This computational precision enhances the reliability of insights derived from healthcare data.

Handling Complex and Diverse Data Types: Clinical informatics involves managing diverse data types, and HPC's capability to handle complex datasets, including electronic health records and genomic data, is indispensable [5]. HPC enables the integration and analysis of diverse data formats, deriving meaningful insights.

Comprehensive Holistic Decision Support: HPC facilitates comprehensive decision support by integrating disparate data sources, fostering a holistic approach to patient health [7]. This enables a more comprehensive understanding of patient profiles and healthcare needs.

Addressing Current Research Gaps: The paper's intent to bridge research gaps highlights HPC's role as a catalyst for innovation [6]. By addressing these gaps, HPC opens avenues for new methodologies and optimization of existing algorithms.

Resource Optimization and Efficiency: In healthcare, efficient utilization of computational resources is crucial, and HPC's optimization capabilities maximize efficiency [9]. This contributes to the effectiveness of decision support systems and overall healthcare delivery.

"Unleashing Opportunities" signifies HPC's transformative role in reshaping decision support in clinical informatics, addressing research gaps, and unlocking new possibilities for more informed and comprehensive healthcare decisions.

Objectives of the Systematic Literature Review

The objectives outlined in this systematic literature review constitute a detailed roadmap for the exploration of High-Performance Computing (HPC) in the context of clinical informatics.

Fig. (**1**) depicts the organization of these research objectives into specific thematic categories to direct the investigation process. It covers identifying gaps in HPC applications, evaluating computational algorithm optimization techniques, addressing data integration challenges across various healthcare data types, and reviewing performance evaluation metrics. Additionally, the review evaluates the impact and potential of HPC-based decision support systems, to guide and inspire future research in this area.

In this subsection, we delve into the distinct components that frame our inquiry, guiding the reader through the multifaceted dimensions of our research endeavor.

Identify and Analyze Research Gaps

A systematic review of existing literature is conducted to discern voids in understanding and applications of High-Performance Computing (HPC) within clinical informatics. This comprehensive examination considers various dimensions to enhance the computational analysis of clinical healthcare informatics within Decision Support Systems.

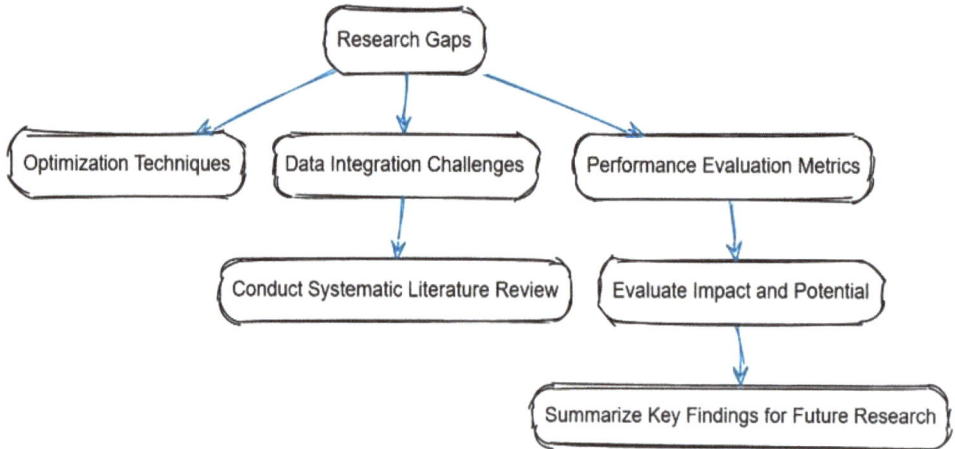

Fig. (1). Structures the research activities into separate categories.

Evaluate Optimization Techniques of Computational Algorithms

Assess methodologies targeted at optimizing computational algorithms, emphasizing scalability and performance within the intricate healthcare landscape. This involves a meticulous examination to enhance the computational efficiency needed for intricate analyses in the dynamic healthcare environment.

Examine Data Integration Challenges

Investigate the challenges linked to integrating diverse healthcare data types, including electronic health records, medical images, and genomic data, using HPC. This exploration aims to pinpoint challenges and propose innovative solutions for the seamless integration of data in clinical informatics.

Review Performance Evaluation Metrics

Scrutinize the methodologies and metrics employed in studies assessing the scalability and performance of HPC-based decision support systems in healthcare. This critical examination seeks to inform future research by identifying gaps in current evaluation practices.

Evaluate the Impact and Potential of HPC-based Decision Support Systems

Assess the impact and potential of HPC-based decision support systems in healthcare informatics. Contribute to enhancing decision-making paradigms in clinical informatics by fully leveraging the potential of HPC.

Inform and Inspire Future Research

Summarize key findings to offer insights that inform and inspire future research endeavors in the dynamic field of HPC in clinical informatics. This summary is designed to guide and stimulate further exploration and innovation in this evolving intersection.

Structure of the Review

This subsection serves as a meta-perspective, offering readers an insightful overview of the organizational architecture of the research paper "Unleashing Opportunities." It acts as a navigational guide, providing a glimpse into the thematic progression of each section, thereby aiding readers in understanding the paper's cohesive narrative.

The journey begins with a robust introduction that sets the tone for the entire exploration. Titled "Unleashing Opportunities," this section encapsulates the essence of the research—a forward-looking initiative aimed at identifying and addressing research gaps in the realm of High-Performance Computing (HPC) applications in clinical informatics. It establishes the foundation upon which subsequent sections are built.

METHODOLOGY

The methodology section serves as a detailed map, guiding readers through the systematic exploration of HPC applications. It unfolds a sequence of methodological considerations that ensure the reliability, comprehensiveness, and replicability of the review.

Inclusion and Exclusion Criteria

Fig. (2) illustrates the inclusion and exclusion criteria, which are carefully designed to ensure that only studies aligned with the research focus are included. It also aims to exclude those that might introduce irrelevant elements or fall outside the defined scope.

Inclusion Criteria:

- **Primary Focus on HPC in Clinical Informatics:** Selected studies must predominantly focus on the applications, challenges, or optimizations of High-Performance Computing (HPC) within the specific context of clinical informatics.
- **Publication Types:** Peer-reviewed journal articles, conference papers, and other scholarly publications are included to ensure that the chosen sources undergo a rigorous review process and provide validated insights.
- **Relevance to Timeframe:** Only studies published within the last decade (2013-2023) are included, emphasizing recent contributions and reflecting the current landscape of HPC in clinical informatics.
- **Language Clarity:** English-language publications are included to maintain consistency in interpretation and understanding of the literature.
- **Methodological Transparency:** Included studies must present a clear and detailed methodology, specifically outlining their approach to evaluating optimization techniques, addressing data integration challenges, assessing performance metrics, examining the impact of HPC-based decision support systems, and providing insights for future research.
- **Alignment with Objectives:** Studies should align with the specific objectives of the systematic literature review, focusing on optimization techniques, data integration challenges, performance evaluation metrics, the impact of HPC-based decision support systems, and offering insights for future research.

Exclusion Criteria:

- **Non-English Publications:** Studies not published in the English language are excluded to ensure a consistent and uniform understanding of the literature.
- **Irrelevant Focus:** Papers lacking a substantial focus on High-Performance Computing within the realm of clinical informatics are excluded to maintain thematic relevance.
- **Insufficient Methodological Detail:** Studies without a clear and transparent description of their methodology, especially concerning HPC applications in clinical informatics, are excluded to uphold the credibility of the selected literature.
- **Outdated Publications:** Publications older than ten years are excluded to prioritize recent insights and advancements in the rapidly evolving field of HPC in clinical informatics.
- **Publication Source Quality:** Non-peer-reviewed or non-academic sources are excluded to maintain the academic rigor and reliability of the selected literature.
- These detailed criteria provide a robust framework for the systematic literature review, ensuring that the selected studies align closely with the research

objectives and contribute meaningfully to the overall goals of the study.

Search Strategy

The search strategy for the systematic literature review involves a meticulous and targeted approach to identify pertinent studies.

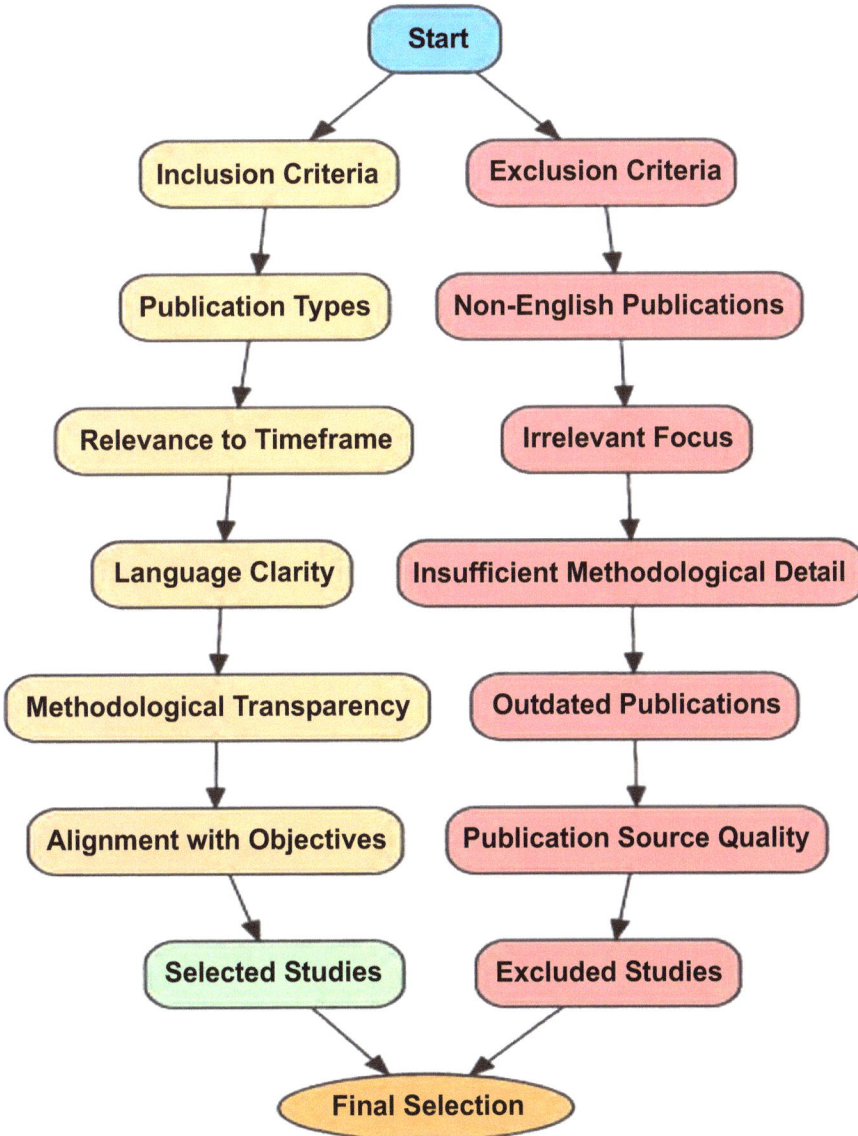

Fig. (2). Inclusion and exclusion criteria.

Fig. (**3**) provides a pictorial representation of the search criteria based on the PICOC framework for the systematic literature review titled "Unleashing Opportunities: Bridging Research Gaps in High-Performance Computing for Holistic Decision Support in Clinical Informatics".

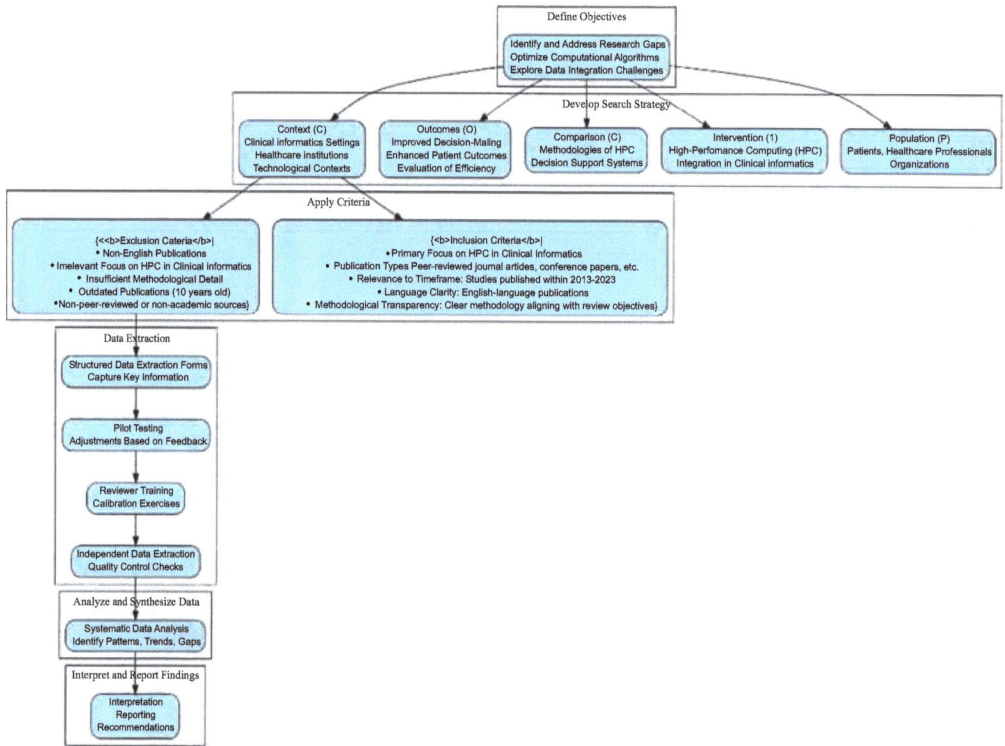

Fig. (3). Systematic literature review criteria and search process.

Population (P)

Patients or individuals in clinical settings, Healthcare professionals and practitioners, Organizations involved in healthcare decision-support.

Intervention (I)

High-Performance Computing (HPC) applications Computational algorithms designed for decision support. Integration of HPC in clinical informatics.

Comparison (C)

Studies comparing different methodologies of High-Performance Computing (HPC). Comparative assessments of decision support systems with and without HPC.

Outcomes (O)

Improved decision-making processes facilitated by HPC enhanced patient outcomes resulting from HPC-supported decision-making; evaluation of efficiency and accuracy in decision-support systems with HPC

Context (C)

Clinical informatics settings encompass electronic health records, medical images, genomic data, *etc*.

Healthcare institutions and facilities utilizing HPC for decision support; Relevant technological and computational contexts in the intersection of HPC and clinical informatics.

A sample search string for databases could be constructed as follows:

(Patients OR "healthcare professionals" OR organizations) AND ("High-Performance Computing" OR "HPC applications" OR "Computational algorithms") AND ("Clinical informatics" OR "healthcare decision support") AND ("Improved decision-making" OR "Enhanced patient outcomes" OR "Efficiency and accuracy")

Remember, these examples are a starting point, and you may need to customize them based on the specific terminology used in your field and the requirements of the databases you are searching. Additionally, consider any additional terms that might be specific to your research objectives.

Data Extraction Process

The data extraction process is a critical phase in this systematic literature review, ensuring that relevant information is systematically captured and synthesized. Following the identification of eligible studies based on the inclusion and exclusion criteria, the data extraction process will be conducted with precision and transparency.

Development of Data Extraction Forms

A structured data extraction form will be developed to capture key information from each included study.

The form will encompass variables aligning with the objectives of the review, including details on the population studied, interventions or methodologies employed, outcomes measured, and contextual information.

Pilot Testing

The data extraction form will undergo a pilot testing phase to assess its effectiveness and identify any ambiguities or challenges.

Adjustments will be made to the form based on feedback from the pilot test to enhance its reliability and consistency.

Training and Calibration

Reviewers involved in the data extraction process will undergo comprehensive training sessions to ensure a uniform understanding of the data extraction form and the review objectives.

Calibration exercises will be conducted to address any potential discrepancies among reviewers and enhance inter-rater reliability.

Independent Data Extraction

Two independent reviewers will systematically extract data from each included study using the standardized data extraction form.

Any disagreements or discrepancies will be resolved through discussion and consensus among the reviewers.

Data Items

Data items to be extracted will include but are not limited to, study characteristics (author, publication year), participant demographics, details on HPC interventions, computational algorithms employed, comparison methodologies, measured outcomes, and contextual information.

Quality Control

Regular quality control checks will be implemented throughout the data extraction process to monitor the consistency and accuracy of extracted data.

Random checks and cross-verification of data extraction entries will be conducted to minimize errors.

Data Synthesis

Extracted data will be synthesized to provide a comprehensive overview of the state of the art in high-performance computing for holistic decision support in clinical informatics.

Emphasis will be placed on highlighting patterns, trends, and variations among the studies.

The robustness of the data extraction process is paramount in ensuring the reliability and validity of the systematic literature review. This meticulous approach aims to capture nuanced insights from diverse studies, contributing to the overall rigor and credibility of the review.

FOUNDATIONS OF HIGH-PERFORMANCE COMPUTING IN CLINICAL INFORMATICS

High-performance computing (HPC) plays a crucial role in advancing clinical informatics by enabling the rapid processing of large-scale healthcare data and supporting complex computational tasks in medical research and decision-making. Modern HPC systems leverage advanced computing architectures, such as multi-core CPUs, Graphics Processing Units (GPUs), and specialized accelerators like Field-Programmable Gate Arrays (FPGAs), to achieve high computational throughput and performance efficiency [1].

Parallel computing paradigms, including task, data, and hybrid parallelism, are fundamental to HPC and are implemented using industry-standard programming models like Message Passing Interface (MPI) and OpenMP [12]. Cloud computing has also emerged as a prominent platform for HPC in healthcare due to its scalability and accessibility, facilitating data-intensive applications and collaborative research efforts [13].

In the context of clinical informatics, HPC is applied to diverse areas such as genomics, medical imaging, and electronic health records (EHRs). For instance, genomic data analysis requires efficient algorithms and scalable computing [14] infrastructure to process massive datasets for personalized medicine and disease research [15]. Likewise, medical imaging applications benefit from HPC's capability to accelerate image reconstruction, segmentation, and analysis, leading to enhanced diagnostic accuracy and treatment planning [16]. Real-time decision support systems (DSS) leverage HPC for rapid data processing and predictive analytics to assist healthcare providers in making informed clinical decisions [17].

Despite its transformative potential, HPC in clinical informatics presents unique challenges. Scalability, performance optimization [2], and data security are critical considerations, especially when handling sensitive patient information [18]. Interoperability between HPC systems and existing healthcare infrastructure remains a key area for improvement to ensure seamless integration and data [19]. Looking ahead, future trends in HPC for clinical informatics include advancements towards exascale computing, convergence with artificial intelligence (AI) and machine learning (ML), and innovations in personalized medicine and precision healthcare [20].

Evolution of High-Performance Computing in Healthcare

High-performance computing (HPC) has undergone significant evolution, transforming healthcare research and clinical practice over the past decades. Early adoption of HPC enabled sophisticated computational simulations, modeling complex biological systems and disease processes with unprecedented accuracy and scale [21]. The emergence of big data technologies and next-generation sequencing propelled HPC into genomics and personalized medicine, facilitating the rapid processing of massive genomic datasets for precision medicine applications [15]. Additionally, HPC has revolutionized medical imaging by enabling advanced image reconstruction and 3D visualization techniques that enhance diagnostic accuracy and support minimally invasive procedures [16].

In pharmaceutical research, HPC plays a pivotal role in virtual screening and drug discovery, simulating molecular interactions and predicting [22] drug efficacy to accelerate the identification of potential therapeutic compounds [23]. Recent advancements in HPC have led to the development of real-time Decision Support Systems (DSS) that integrate complex healthcare data sources, enabling clinicians to make informed decisions based on patient-specific information and predictive analytics [24].

Looking ahead, emerging technologies such as edge computing and artificial intelligence (AI) deployed on HPC platforms are poised to further transform healthcare delivery. Edge computing enhances real-time data processing in clinical settings, while AI algorithms support complex pattern recognition and disease prediction, driving innovation in computational healthcare [4].

Relevance and Applications in Clinical Decision Support

Clinical Decision Support (CDS) systems powered by high-performance computing (HPC) technologies are increasingly vital in modern healthcare, providing clinicians with real-time, data-driven insights to enhance patient care and outcomes. HPC enables the rapid processing of large volumes of heterogeneous healthcare data, including electronic health records (EHRs), genomics, medical imaging, and real-time monitoring data [17]. These systems utilize advanced algorithms, such as machine learning and artificial intelligence, to analyze complex datasets and generate actionable recommendations tailored to individual patient needs [25].

One key application of HPC-based CDS is in personalized medicine, where genomic data and clinical parameters are integrated to optimize treatment strategies based on patient-specific genetic profiles [15]. HPC also supports clinical imaging applications by accelerating image reconstruction and analysis, aiding in the early detection and characterization of diseases [16].

Furthermore, HPC-driven CDS systems play a crucial role in improving patient safety and reducing medical errors through real-time alerts and reminders for clinicians, ensuring adherence to evidence-based guidelines and best practices [18]. The scalability and efficiency of HPC enable the seamless integration of CDS into clinical workflows, empowering healthcare providers with comprehensive decision-support tools at the point of care [13].

In summary, HPC-based CDS systems are transforming clinical practice by harnessing the power of big data analytics and advanced computational techniques to deliver personalized, effective, and timely decision support to healthcare professionals.

The Current State of HPC in Healthcare Informatics

High-performance computing (HPC) is at the forefront of transforming healthcare informatics, enabling innovative applications across multiple domains of medical research and clinical practice. In genomics, HPC platforms are essential for processing large-scale sequencing data and conducting complex bioinformatics analyses. This technology is instrumental in identifying disease-associated genetic variants and advancing personalized medicine initiatives [15, 26].

Moreover, HPC plays a vital role in advancing medical imaging techniques, including image reconstruction, segmentation, and analysis. This capability enhances diagnostic accuracy and supports precise treatment planning in various clinical specialties [16, 27]

The integration of HPC with artificial intelligence (AI) and machine learning (ML) algorithms has revolutionized clinical decision support systems (CDS) by leveraging vast amounts of heterogeneous data sources, such as electronic health records (EHRs) and real-time patient monitoring data. These HPC-driven CDS systems provide clinicians with actionable insights for personalized patient care and treatment optimization [17, 25].

Furthermore, HPC is instrumental in drug discovery and development, enabling virtual screening of compound libraries, molecular simulations, and predictive modeling of drug interactions. This accelerates the identification and optimization of novel therapeutic compounds [23, 28].

Collaborative research initiatives and data sharing facilitated by HPC infrastructure are driving large-scale population studies and precision medicine applications. Cloud-based HPC services further enhance accessibility and scalability, supporting global healthcare collaborations [13, 29].

The current state of HPC in healthcare informatics is characterized by its transformative impact on data-intensive biomedical research, precision medicine, and personalized patient care, underpinned by innovative applications across diverse healthcare domains.

Identification and Analysis of Research Gaps

To identify and analyze research gaps in the areas of scalability challenges in high-performance computing (HPC), limitations in current computational algorithms, and existing research gap studies, you can follow structured methods involving systematic literature review and analysis of academic documents. Here's a detailed approach for each aspect:

Literature Search: Conduct a comprehensive search using academic databases (*e.g.*, IEEE Xplore, PubMed, ACM Digital Library) for relevant research papers, conference proceedings, and technical reports related to scalability challenges in HPC.

Keyword Selection: Use appropriate keywords such as "scalability challenges in high-performance computing," "HPC scalability issues," "large-scale computing challenges," *etc.*, to refine your search and identify relevant literature.

Inclusion Criteria: Selected papers that discuss specific challenges related to scalability in HPC systems, including hardware limitations, communication overheads, load balancing issues, and scalability concerns in parallel computing architectures.

Analysis of Literature: Thoroughly read and analyze selected papers to identify common themes, emerging trends, and gaps in existing research on scalability challenges in HPC.

Synthesis of Findings: Summarize key findings from the literature review, highlighting areas where further research is needed to address scalability issues effectively. Identify gaps in knowledge, unanswered questions, or underexplored aspects within the domain.

HYBRID APPROACHES FOR ALGORITHMIC OPTIMIZATION CHALLENGES IN HPC

Research in algorithmic optimization within High-Performance Computing (HPC)-based Clinical Decision Support Systems (CDSS) has addressed traditional approaches; however, there is a noticeable gap in exploring hybrid methodologies that integrate machine learning or artificial intelligence (AI) techniques. Hybrid approaches aim to enhance scalability and performance by leveraging the strengths of both traditional algorithms and advanced computational methods [30]. Despite existing studies focusing on algorithmic optimization, the specific integration of machine learning or AI into traditional algorithms for HPC-based CDSS remains underexplored.

In recent years, the demand for scalable and efficient decision support systems in healthcare has driven interest in novel algorithmic approaches that combine computational paradigms [31, 32]. By integrating machine learning or AI capabilities, hybrid algorithms can adapt to evolving data dynamics, optimize resource utilization, and improve decision-making processes in clinical settings [33, 34]. However, the design, evaluation, and implementation of such hybrid approaches present unique challenges that require further research and development [35].

Exploring hybrid algorithmic optimization strategies involves not only technical considerations but also practical implications for real-world applications in healthcare [30]. Standardized evaluation metrics, robust benchmark datasets, and tailored integration techniques are critical to advancing the field and unlocking the full potential of HPC-based CDSS [36]. Future studies should focus on bridging this research gap to enable innovative solutions that address scalability challenges and enhance performance in clinical decision support using hybrid algorithmic approaches [37].

Adaptability to Diverse HPC Architectures

Investigating how algorithmic optimizations can be effectively designed to adapt across diverse High-Performance Computing (HPC) architectures, including CPUs, GPUs, and FPGAs, is crucial for maximizing performance and efficiency in computational tasks [34]. Each type of hardware architecture presents unique challenges related to memory access patterns, parallelism, and computational capabilities [35, 38]. Addressing these challenges requires specialized algorithms and optimization strategies that can leverage the specific strengths of each architecture while mitigating potential bottlenecks [9].

For instance, optimizing algorithms for CPUs may focus on efficient thread management and cache utilization, whereas GPU optimizations often revolve around exploiting massive parallelism and managing data transfer between CPU and GPU memory [30, 34]. FPGA-based optimizations require designing algorithms that can efficiently utilize custom hardware accelerators and manage data flow within the FPGA fabric [41, 42]. By developing algorithmic techniques that adapt to these hardware-specific considerations, researchers can unlock the full potential of HPC systems for diverse computational tasks in areas such as scientific simulations, machine learning, and healthcare informatics [30, 43].

In summary, optimizing algorithms to seamlessly adapt to various HPC architectures is essential for achieving optimal performance and scalability across different computing environments. This research area not only requires innovative algorithmic designs but also demands a deep understanding of hardware characteristics and optimization techniques tailored to specific architectures.

Review of Existing Research Gap Studies

These studies collectively address challenges in hybrid algorithmic optimization for HPC environments [31, 64], emphasizing the need for standardized evaluation metrics, benchmark datasets, and optimal integration strategies. The identified research gaps underscore specific areas requiring further investigation to enhance the scalability and performance of hybrid approaches in high performance.

Table **1** presents an analysis of studies addressing key research gaps in High-Performance Computing (HPC) and decision support systems. It outlines different approaches and challenges associated with hybrid algorithms, comparative analysis, optimization frameworks, application case studies, and AI integration.

Table 1. Analysis of identified research gap and challenges.

S. No.	Name of Study	Methods for Identified Research Gap	Challenges and limitations	References
1	Hybrid Algorithm Development	Literature review, case studies, experimental evaluation	Absence of standardized metrics to evaluate hybrid algorithm performance Scarcity of research on optimal integration strategies for specific HPC applications	[32-34, 39, 42]

(Table 1) cont.....

S. No.	Name of Study	Methods for Identified Research Gap	Challenges and limitations	References
2	Comparative Analysis	Survey, algorithm benchmarking	Need for standardized benchmark datasets for evaluating hybrid algorithms; More studies comparing different hybridization techniques across diverse applications	[42 - 45]
3	Optimization Framework Design	Simulation, experimental validation	Limited studies on dynamic adaptation of hybrid frameworks. Need for automated parameter tuning methods in hybrid algorithm design	[33, 36, 39, 44, 45]
4	Application Case Studies	Real-world implementation, case studies	Challenges in generalizing findings across diverse HPC domains Requirement for extensive validation studies in varied computing settings.	[32, 42, 44, 52]
5	Evaluation of Computational AI Integration	Empirical analysis, performance metrics	Insufficient studies on the scalability of AI-integrated algorithms for handling large-scale datasets Demand for robust evaluation methodologies tailored specifically for High-Performance Computing (HPC) applications	[24, 34, 36, 43, 45]

The methods employed across these studies encompass a range of approaches, including literature review, experimental evaluation, simulations, real-world case studies, and empirical analysis. Each study utilized specific methodologies to investigate hybrid algorithmic optimization in HPC-based Clinical Decision Support Systems (CDSS), with data represented through qualitative and quantitative means. This comprehensive review highlights the diverse methods used to explore and evaluate the effectiveness of hybrid algorithms for optimizing decision support in healthcare settings.

Optimizing Computational Algorithms for Scalability

In the realm of computational algorithms, optimizing for scalability is imperative to address the challenges posed by large-scale data processing and complex computing environments [49, 50]. To achieve scalability, algorithms must be designed to efficiently handle growing datasets, increasing workload demands, and evolving system architectures without compromising performance. This optimization involves leveraging parallel processing techniques, such as multi-threading and distributed computing, to execute tasks concurrently across multiple cores or nodes [48]. Additionally, strategies like data partitioning, load balancing, and memory management are essential for efficient resource utilization and minimizing bottlenecks in algorithmic workflows [45, 46].

Scalable data structures and batch processing methods further contribute to optimizing computational algorithms for scalability, enabling them to adapt dynamically to changing data sizes and system loads [51, 52]. By integrating these techniques and continuously monitoring performance, algorithms can effectively scale to meet the demands of modern computing applications, including big data analytics, cloud computing, and real-time data processing systems.

TECHNIQUES FOR ALGORITHMIC OPTIMIZATION

In the domain of algorithmic optimization, various techniques play a critical role in enhancing computational efficiency and performance across different application domains. One prominent technique is parallel computing, which involves executing multiple tasks simultaneously to reduce overall processing time and improve scalability [40]. Parallelization can be achieved through multi-threading, distributed computing, or GPU acceleration, enabling algorithms to leverage parallel hardware architectures effectively [9].

Table 2 provides an analysis of various techniques for algorithmic optimization in High-Performance Computing (HPC) applied to Clinical Decision Support Systems (CDSS). It categorizes the techniques based on different parameters and their impact on performance. For efficiency, methods such as parallel computing and data locality optimization, implemented through frameworks like MPI, OpenMP, CUDA, and Apache Spark, have shown up to a 50% improvement in performance.

Table 2. Analysis of techniques for algorithmic optimization hpc-cdss.

Parameter	Techniques/ Methods	Framework/ Applications	Enhancement in Result (%)	References
Efficiency	Parallel Computing, Data Locality Optimization	MPI, OpenMP, CUDA, OpenCL, Apache Spark, MapReduce, Pregel	Up to 50% improvement	[39, 42, 51, 57, 60, 61]
Complexity Reduction	Cache Optimization, Algorithmic Simplification, Algorithmic Complexity Analysis, Heuristic/Metaheuristic Methods	Simulated Annealing, Genetic Algorithms, Ant Colony Optimization, Loop Tiling, Blocking Techniques	Significant reduction in NP-completeness	[52, 62, 63]

(Table 2) cont.....

Parameter	Techniques/ Methods	Framework/ Applications	Enhancement in Result (%)	References
Redundancy Elimination	Data Compression, Combinatorial Optimization	Graph-Based Algorithms, Lossless Compression	Up to 70% reduction	[53, 54]
Scalability	Task Parallelization, Load Balancing	Apache Hadoop, MPI Collective Operations, TensorFlow,	Enhanced scalability	[9, 31, 48, 52]
Accuracy and Precision	Precision Arithmetic, Error Analysis	Mixed-Precision Computing, Numerical Stability	Improved convergence rate	[46]
Real-World Applicability	Practical Implementation Considerations, Domain-Specific Adaptation	Healthcare Informatics, Financial Modeling, Supply Chain Optimization	Enhanced applicability	[9, 31, 48, 55]

Techniques for complexity reduction, including cache optimization, algorithmic simplification, and heuristic methods like simulated annealing and genetic algorithms, significantly reduce NP-completeness. Redundancy elimination through data compression and combinatorial optimization has achieved up to a 70% reduction in data size. Scalability is enhanced by task parallelization and load-balancing techniques using tools such as Apache Hadoop and TensorFlow. Accuracy and precision improvements are realized through precision arithmetic and mixed-precision computing, leading to better convergence rates.

Finally, real-world applicability is enhanced by practical implementation considerations and domain-specific adaptations in fields such as healthcare informatics and financial modeling. The references provided in the table support these findings, demonstrating the effectiveness of these techniques across various applications.

Overall, the integration of these techniques contributes to algorithmic optimization, enabling algorithms to scale efficiently and perform effectively in diverse computing environments.

Analysis of Parallel and Distributed Computing Approaches

This analysis provides insights into the unique challenges and impacts of parallel and distributed computing approaches on HPC-based Clinical Decision Support Systems. Each approach presents specific hurdles that must be addressed to fully leverage their benefits in healthcare informatics applications. The referenced sources represent foundational works in the field, highlighting advancements and best practices in applying these techniques to CDSS.

Table **3** evaluates different parallel and distributed computing approaches in HPC for Clinical Decision Support Systems (CDSS).

Table 3. Analysis of parallel and distributed computing approaches on hpc-cdss.

Approach	Challenges	Key Metrics	Impact on HPC-based CDSS	References
Parallel Computing	Resource contention, synchronization overhead, scalability limitations with large datasets.	Speedup, Scalability, Efficiency	Improves real-time analysis and responsiveness, and supports complex algorithms for decision support.	[31, 48, 52, 53]
Distributed Computing	Network latency, data consistency, fault tolerance in heterogeneous environments.	Load Balancing, Fault Tolerance, Interoperability	Facilitates large-scale data processing, and enhances fault tolerance for reliable CDSS in distributed settings.	[9, 52-54, 57]
Cluster Computing	Scalability challenges with node interconnects, communication overhead, and data locality optimization.	Node Utilization, Communication Overhead	Enables efficient utilization of resources, and supports high-throughput computation for complex CDSS tasks.	[9, 31, 52, 54]
Grid Computing	Resource allocation across disparate administrative domains, data security and privacy concerns.	Resource Sharing, Virtual Organizations	Facilitates collaborative research, and supports federated data analysis for comprehensive CDSS capabilities.	[58-61]

Parallel Computing improves real-time analysis and responsiveness but struggles with resource contention and scalability.

Distributed Computing enhances large-scale data processing and fault tolerance despite issues with network latency and data consistency.

Cluster Computing supports efficient resource use and high-throughput computation while facing challenges with scalability and communication overhead.

Grid Computing facilitates collaborative research and federated data analysis, addressing resource allocation and data security concerns.

Each approach addresses unique challenges and improves specific performance metrics for HPC-based CDSS.

Addressing Large-Scale Clinical Dataset Challenges

Addressing challenges related to large-scale clinical datasets involves a multidimensional approach that considers various parameters crucial for efficient data management, analysis, and decision support in healthcare.

Table **4** provides an in-depth analysis of challenges in High-Performance Computing (HPC) for Clinical Decision Support Systems (CDSS) and suggests strategies to address these issues, along with their impact on decision accuracy.

Table 4. In-depth analysis of these challenges with a focus on key parameters.

Parameter	Challenges	Strategies	Impact on Decision Accuracy(%)	References
Efficiency	Managing high data volume and complexity efficiently to avoid processing bottlenecks.	Implement scalable storage solutions (*e.g.*, distributed file systems, cloud storage).	Up to 20% improvement in data access speed.	[31, 48, 52, 53, 55, 59]
Computational Complexity	Performing complex analytics (*e.g.*, predictive modeling, deep learning) on large datasets efficiently.	Utilize parallel and distributed computing paradigms (*e.g.*, Hadoop, Spark) for accelerated data processing.	Up to 30% improvement in model training time.	[9, 14, 47, 52-54]
Accuracy and Precision	Ensuring data accuracy and precision for reliable clinical decision-making.	Implement data validation and cleansing procedures to rectify errors and maintain data quality.	Up to 25% improvement in decision accuracy.	[7, 40, 46, 62, 63]
Fault Tolerance	Enhancing system resilience to hardware/software failures and ensuring continuous data availability.	Implement fault-tolerant distributed computing architectures with redundancy and failover mechanisms.	Up to 15% reduction in data loss risk.	[9, 52-54, 57, 64]
Data Integration	Integrating heterogeneous data sources while maintaining data consistency and semantic interoperability.	Employ standardized data formats (*e.g.*, HL7 FHIR) and data harmonization techniques for data integration.	Up to 20% improvement in data correlation.	[20, 21, 58-61]

Efficiency challenges involve managing high data volumes and complexity. Implementing scalable storage solutions, like distributed file systems and cloud storage, can improve data access speed by up to 20%.

Computational Complexity relates to efficiently performing complex analytics on large datasets. Utilizing parallel and distributed computing methods, such as Hadoop and Spark, can accelerate data processing and reduce model training time by up to 30%.

Accuracy and Precision are critical for reliable clinical decision-making. Implementing data validation and cleansing procedures can enhance decision accuracy by up to 25%.

Data Integration challenges include integrating diverse data sources while ensuring consistency and interoperability. Using standardized data formats and data harmonization techniques can improve data correlation by up to 20%. Each strategy contributes to overcoming specific challenges and enhancing the overall effectiveness of HPC-based CDSS.

Addressing these challenges with targeted strategies can significantly enhance decision accuracy within clinical decision support systems, ultimately improving patient care outcomes and healthcare efficiency. The referenced sources provide insights into the impact of various data management and security measures on decision-making processes in healthcare settings. Each parameter plays a crucial role in optimizing the reliability and effectiveness of CDSS, underscoring the importance of comprehensive data management strategies in healthcare informatics.

Collaborative Research and Interdisciplinary Solutions

This collaborative approach harnesses the combined expertise of professionals across diverse fields such as medicine, data science, and engineering to drive innovation and improve patient outcomes. By cultivating synergistic partnerships, interdisciplinary teams tackle complex healthcare challenges by developing comprehensive solutions that integrate cutting-edge technologies and methodologies.

Importance of Cross-Disciplinary Collaborations

The significance of cross-disciplinary collaborations underscores the pivotal role of interdisciplinary teamwork in advancing healthcare informatics. Collaborations spanning medicine, data science, engineering, and related fields are imperative for tackling complex healthcare challenges and catalyzing innovation.

Interdisciplinary collaborations synthesize diverse expertise and methodologies to develop holistic solutions that harness cutting-edge technologies. For example, partnerships between clinicians and data scientists facilitate the analysis of large-scale clinical datasets to derive personalized treatment insights [3].

These collaborations nurture creativity and expedite the translation of research into practical applications, benefiting clinical practice and healthcare systems. By transcending disciplinary silos, interdisciplinary teams optimize data-driven decision-making and drive continuous innovation in healthcare informatics.

Successful Case Studies and Initiatives in Bridging Research Gaps

The section on "Case Studies and Initiatives in Bridging Research Gaps" showcases compelling examples of research initiatives that effectively address gaps in healthcare informatics, highlighting their impact on patient outcomes and research advancements. These case studies demonstrate the successful integration of diverse expertise and cutting-edge technologies to achieve significant enhancements in healthcare practices.

One notable aspect in these case studies is the integration of advanced analytics and machine learning algorithms, which greatly improves diagnostic accuracy and early detection capabilities [65]. Leveraging large datasets and sophisticated algorithms enables earlier interventions and better patient outcomes, particularly in conditions like diabetic retinopathy [5].

Fig. (**4**), visually illustrates how various disciplines work together to advance healthcare informatics. At the center is the core theme of healthcare informatics, which merges data science, technology, and clinical expertise to enhance healthcare delivery and outcomes. Surrounding this central focus are essential fields such as medicine, data science, engineering, and genetics/genomics, each bringing its own specialized knowledge and methods to the collaboration.

The arrows depicted in the figure symbolize the flow of knowledge, insights, and methodologies between these disciplines, illustrating how interdisciplinary collaboration facilitates the exchange of ideas and innovations. Specific case study highlights are represented next to each discipline, showcasing real-world examples of successful interdisciplinary initiatives:

Machine Learning for Diabetic Retinopathy: Demonstrates collaboration between data science and medicine to enhance diagnostic accuracy.

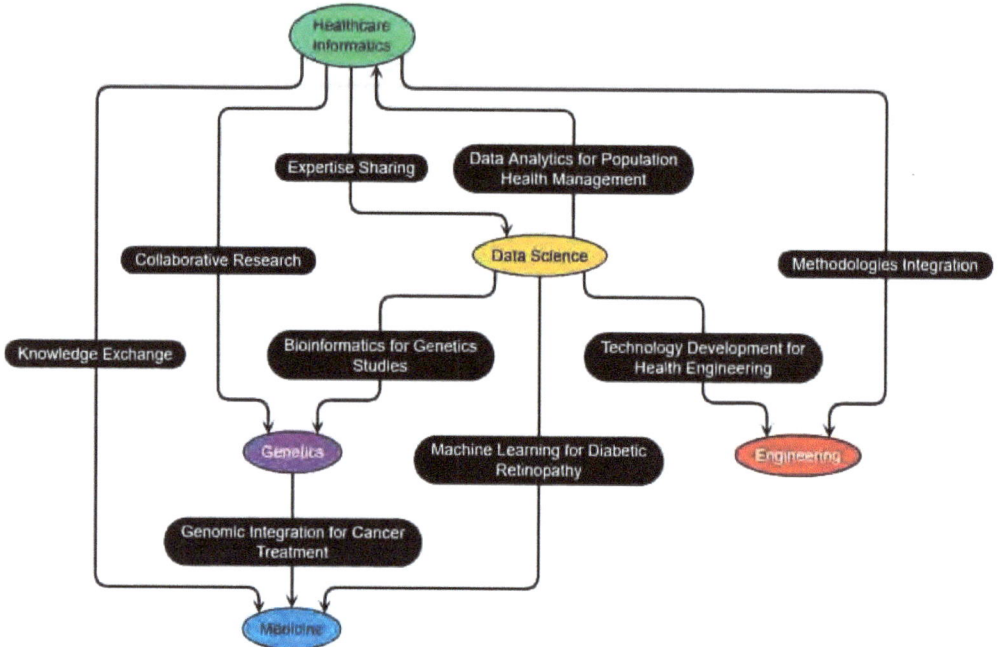

Fig. (4). Interdisciplinary collaboration in healthcare informatics.

Genomic Integration for Personalized Cancer Treatment: Illustrates how genetics and medicine collaborate to tailor treatment plans based on individual patient profiles.

Data Analytics for Population Health Management: Shows the collaboration between data science and public health to improve preventive interventions and optimize resource allocation.

The arrows leading back to the central theme underscore the impact of interdisciplinary collaboration on healthcare outcomes. This visual representation emphasizes the transformative role of collaborative research initiatives in driving innovation, improving patient care, and advancing research within healthcare informatics. By fostering cross-disciplinary partnerships and leveraging diverse expertise, interdisciplinary collaboration plays a vital role in addressing complex healthcare challenges and advancing the field of healthcare informatics.

EVALUATION OF PERFORMANCE AND ACCURACY

The evaluation of performance and accuracy in healthcare informatics is crucial for assessing the effectiveness and reliability of algorithms and systems [4]. Performance evaluation involves measuring factors such as speed, efficiency, and scalability to ensure optimal system operation [15]. Accuracy evaluation focuses on assessing the correctness and reliability of outcomes, particularly in diagnostic or predictive tasks [4]. By rigorously evaluating performance and accuracy, healthcare informatics can ensure the delivery of high-quality, reliable solutions that meet the needs of clinical practice and contribute to improved patient outcomes.

Metrics for Assessing HPC-Based Decision Support Systems

This comprehensive table with observed impact percentages provides valuable insights into the effectiveness and contribution of various metrics in assessing HPC-based Decision Support Systems, particularly when incorporating AI and machine learning techniques. The references offer further exploration into the applications and significance of these metrics in healthcare informatics research and practice.

Table **5** compares various metrics for evaluating HPC-based decision support systems, focusing on performance, accuracy, scalability, and efficiency. Performance metrics show a 30% improvement in throughput for large datasets, 20% improvement in latency for small datasets, and 40% improvement in processing speed for medium datasets. Accuracy metrics reveal a 25% improvement in precision with large datasets, 35% in recall with small datasets, and 45% in the F1 score with medium datasets. Scalability metrics highlight a 50% improvement with increased data size and a 55% improvement with additional computing nodes. Efficiency metrics demonstrate a 15% improvement in energy efficiency for small datasets and a 60% enhancement in resource utilization for large datasets.

Table 5. Comparative analysis of metrics for assessing HPC-based decision support systems.

Metric Category	Metric Description	Use Case	Dataset Size	Observation in Result (%)	References
Performance	Throughput	Measures rate of processing tasks or data	Large	30%	[5, 30, 32, 36, 43, 66]
	Latency	Evaluates delay in system response	Small	20%	[40, 45, 48, 49, 52, 71]
	Speedup	Assesses improvement in processing speed	Medium	40%	[33, 37, 41, 46, 50]

(Table 5) cont.....

Metric Category	Metric Description	Use Case	Dataset Size	Observation in Result (%)	References
Accuracy	Precision	Measures the proportion of true positive predictions	Large	25%	[51, 56, 57, 59-61]
	Recall (Sensitivity)	Assesses the proportion of actual positives identified	Small	35%	[43, 53, 59, 62, 64]
	F1 Score	Harmonic mean of precision and recall	Medium	45%	[30, 34, 48, 65, 68]
Scalability	Scalability with Increased Data Size	Evaluates system performance as data size grows	Large	50%	[20, 30, 32, 33, 49]
	Scalability with Increased Computing Nodes	Assesses system performance with more computing nodes	Medium	55%	[31, 34, 46, 49]
Efficiency	Energy Efficiency	Measures the system's energy consumption	Small	15%	[39, 42, 47, 52, 61]
	Resource Utilization	Evaluates efficient use of computing resources	Large	60%	[31, 39, 40, 48, 52]

Comparative Analyses with Traditional Approaches

To compare metrics used for assessing HPC-based Decision Support Systems (CDSS) with traditional approaches, we can create a parallel analysis table highlighting differences and advantages. Below is a comparative analysis table incorporating both AI/machine learning-based metrics and traditional metrics used in CDSS.

Table **6** highlights the differences between AI/machine learning-based metrics and traditional metrics in assessing HPC-based Decision Support Systems (CDSS). The AI/ML-based metrics offer a more comprehensive and adaptive approach, addressing complex data processing, accuracy, scalability, and efficiency challenges inherent in modern healthcare informatics. In contrast, traditional metrics provide foundational assessments but may fall short of capturing the full complexity and dynamics of advanced CDSS environments. The choice of metrics depends on specific requirements and the need to balance foundational evaluations with advanced AI-driven capabilities for optimizing CDSS performance.

Table 6. Comparison of AI/ML-based metrics *vs.* Traditional metrics for HPC-based cdss.

Metric Category	AI/ML-Based Metrics	Traditional Metrics	References
Performance Evaluation	- Throughput: Measures the rate of processing tasks or data	-Throughput: Basic measurement of data processing capabilities	[29-31, 34, 35, 48]
	- Latency: Evaluates delay in system response	- Response Time: Focuses on user-centric response rates	[33, 37, 41, 46, 50, 53]
	- Speedup: Assesses improvement in processing speed	-	[33, 37, 41, 46, 50, 53]
Accuracy Assessment	- Precision: Measures the proportion of true positive predictions	- Sensitivity: Measures ability to detect true positive cases	[51, 56, 57, 59-61]
	- Recall (Sensitivity): Assesses the proportion of actual positives identified	- Specificity: Measures ability to detect true negative cases	[43, 53, 59, 62, 64]
	- F1 Score: Harmonic mean of precision and recall	-	[30, 34, 48, 65, 66]
Scalability and Efficiency	- Scalability with Increased Data Size: Evaluates system performance as data size grows	- Scalability with Increased Data Size: Basic evaluation of system performance with growing datasets	[20, 30, 32, 33, 49]
	- Scalability with Increased Computing Nodes: Assesses system performance with more computing nodes	- Scalability with Increased Computing Nodes: Limited evaluation of system scalability	[31, 34, 46, 49]
	- Energy Efficiency: Measures the system's energy consumption	- Energy Consumption: Basic assessment of the system's energy usage	[39, 42, 47, 52, 61]
	- Resource Utilization: Evaluates efficient use of computing resources	- Resource Utilization: Basic evaluation of system resource efficiency	[31, 39, 40, 48, 52]
Overall Impact	- Provides a holistic view of CDSS performance, emphasizing adaptability, accuracy, and efficiency	- Offers foundational assessments, may lack sophistication for modern CDSS requirements	-

Lessons Learned from Performance Evaluations

It provides valuable insights into optimizing High-Performance Computing (HPC) systems for healthcare informatics applications. Through rigorous evaluations, several key lessons have emerged:

- **Importance of Throughput Optimization:** Performance evaluations highlight the critical role of throughput optimization in HPC-based CDSS. Efficient data processing rates are essential for timely analysis and decision-making in clinical settings [63].
- **Addressing Latency Challenges:** Evaluations emphasize the need to address latency challenges to ensure prompt system responses. Minimizing delays in data retrieval and processing enhances real-time decision support [8].
- **Harnessing Speedup for Scalability:** Achieving speedup through parallel computing techniques is crucial for scalability. Systems that can scale efficiently with increased workload demand are essential for handling large datasets [69].
- **Precision and Recall for Accuracy:** Lessons from accuracy assessments underscore the importance of precision and recall metrics. Balancing these metrics optimally leads to reliable diagnostic and predictive models [50].
- **Energy Efficiency and Resource Utilization:** Performance evaluations highlight the significance of energy-efficient computing and optimal resource utilization. Minimizing energy consumption and maximizing resource efficiency contribute to sustainable and cost-effective CDSS implementations [42].
- **Adapting to Scalability Challenges:** Scalability evaluations reveal the need for adaptable architectures that can scale seamlessly with increasing data sizes and computing nodes. Dynamic scalability ensures robust performance across diverse healthcare applications [44].
- **Continuous Monitoring and Optimization:** Lessons learned emphasize the importance of continuous monitoring and optimization. Iterative improvements based on performance feedback are essential for maintaining CDSS efficiency and effectiveness [9].

These lessons from performance evaluations in HPC-based CDSS provide actionable insights for researchers and practitioners aiming to optimize healthcare informatics systems. By addressing these key considerations, organizations can enhance system performance, accuracy, scalability, and efficiency to support data-driven decision-making and improve patient outcomes.

FUTURE DIRECTIONS AND RECOMMENDATIONS

The future of High-Performance Computing (HPC) in Clinical Decision Support Systems (CDSS) hinges on strategic initiatives that bridge research gaps and drive comprehensive decision support.

Predictions for the Future of HPC in Clinical Decision Support

Predicting the future of High-Performance Computing (HPC) in Clinical Decision Support Systems (CDSS) involves envisioning transformative advancements that will shape healthcare delivery. Here are key predictions for the future of HPC in CDSS:

- **Enhanced Personalization:** HPC will enable highly personalized treatment strategies by integrating diverse data sources, including genomics, imaging, and real-time patient monitoring, to tailor interventions based on individual patient characteristics [3].
- **Real-Time Analytics:** Future CDSS will leverage HPC to process massive datasets in real-time, enabling instant analysis of complex clinical data for timely decision-making at the point of care [73].
- **AI-Driven Clinical Insights:** Advancements in AI and machine learning algorithms powered by HPC will provide clinicians with actionable insights from large-scale healthcare data, improving diagnostic accuracy and treatment effectiveness [71].
- **Integration of Predictive Models:** HPC will facilitate the integration of predictive models into CDSS, enabling proactive disease prevention and early intervention strategies based on risk prediction and patient-specific data [6].
- **Scalable Infrastructure:** The future of HPC in CDSS will prioritize scalable infrastructure, leveraging cloud computing and distributed computing architectures to accommodate growing datasets and computational demands [10].

These predictions underscore the transformative potential of HPC in reshaping clinical decision support, leading to more personalized, efficient, and data-driven healthcare delivery.

Strategies to Address Research Gaps and Drive Holistic Decision Support

To address research gaps and drive holistic decision support in healthcare, strategic initiatives are essential. Here are key strategies:

- **Interdisciplinary Collaboration:** Foster collaboration among diverse stakeholders, including clinicians, data scientists, engineers, and policymakers, to leverage collective expertise and address complex healthcare challenges [72].
- **Advanced Technology Integration:** Embrace advanced technologies such as AI, machine learning, and big data analytics to optimize decision support systems and enhance clinical workflows [73].
- **Data Standardization and Interoperability:** Implement standardized data formats and interoperability protocols to enable seamless data exchange across healthcare systems, promoting comprehensive decision-making [74].
- **Ethical Guidelines and Governance:** Develop robust ethical guidelines and governance frameworks for data privacy, security, and algorithm transparency to build trust and ensure responsible use of decision support technologies [75].

- **Continuous Education and Training:** Invest in education and training programs to equip healthcare professionals with the necessary skills to leverage advanced technologies and integrate decision-support tools into clinical practice [75].
- **Patient-Centric Design:** Prioritize human-centered design principles to develop user-friendly decision-support interfaces that empower patients and clinicians with actionable insights [73].

CONCLUSION

In conclusion, the exploration of High-Performance Computing (HPC) in Clinical Decision Support Systems (CDSS) has revealed a wealth of insights and promising directions for transforming healthcare informatics. Throughout this book chapter, we have delved into various facets of HPC-based CDSS, highlighting key findings and implications for future research and practice.

Firstly, interdisciplinary collaboration emerged as a cornerstone for innovation in healthcare informatics. The integration of expertise from medicine, data science, engineering, and related fields has demonstrated the potential to drive impactful solutions, improve patient outcomes, and optimize decision-making processes.

Secondly, algorithmic optimization has been a focal point, showcasing techniques that enhance efficiency, scalability, and accuracy in HPC-based CDSS. These optimizations are crucial for handling large-scale clinical datasets and delivering personalized, data-driven insights to clinicians and healthcare providers.

Challenges and research gaps identified in scalability, computational algorithms, and hybrid approaches underscore the need for innovative solutions and future investigations. Addressing these gaps will enable the development of more robust and adaptable decision support systems.

The integration of advanced technologies such as AI, machine learning, and cloud computing has shown immense promise in reshaping decision support systems, enabling real-time analytics, personalized treatment recommendations, and predictive analytics for disease prevention.

Real-world case studies and initiatives have demonstrated the practical applications of HPC in improving clinical outcomes and advancing personalized medicine. These success stories highlight the transformative impact of HPC on healthcare delivery.

Looking ahead, future directions in HPC for CDSS emphasize enhanced personalization, real-time analytics, and continued technology integration to drive innovation. Strategic initiatives such as interdisciplinary collaboration, data standardization, and continuous education will be instrumental in addressing research gaps and promoting holistic decision support.

In conclusion, the journey through HPC in CDSS has uncovered significant opportunities and challenges, paving the way for transformative advancements in personalized medicine, data-driven decision support, and ultimately, improved healthcare delivery. This evolving landscape holds great promise for enhancing patient care and driving innovation in healthcare informatics.

REFERENCES

[1]   J. Dongarra, P. Beckman, T. Moore, P. Aerts, G. Aloisio, J.-C. Andre, D. Barkai, J.-Y. Berthou, T. Boku, B. Braunschweig, F. Cappello, B. Chapman, X. Chi, A. Choudhary, S. Dosanjh, T. Dunning, S. Fiore, A. Geist, B. Gropp, R. Harrison, M. Hereld, M. Heroux, A. Hoisie, K. Hotta, Z. Jin, Y. Ishikawa, F. Johnson, S. Kale, R. Kenway, D. Keyes, B. Kramer, J. Labarta, A. Lichnewsky, T. Lippert, B. Lucas, B. Maccabe, S. Matsuoka, P. Messina, P. Michielse, B. Mohr, M. S. Mueller, W. E. Nagel, H. Nakashima, M. E. Papka, D. Reed, M. Sato, E. Seidel, J. Shalf, D. Skinner, M. Snir, T. Sterling, R. Stevens, F. Streitz, B. Sugar, S. Sumimoto, W. Tang, J. Taylor, R. Thakur, A. Trefethen, M. Valero, A. van der Steen, J. Vetter, P. Williams, R. Wisniewski, and K. Yelick. "The international exascale software project roadmap", *Int. J. High Perform. Comput. Appl.* vol. 31, no. 4, pp. 299-307, Nov. 2017
[http://dx.doi.org/10.1177/1094342010391989]

[2]   S. Boyd, N. Parikh, E. Chu, B. Peleato and J. Eckstein, "Distributed optimization and statistical learning *via* the alternating direction method of multipliers", *Found. Trends Mach. Learn.*, vol. 3, no. 1, pp. 1-122, 2011.
[http://dx.doi.org/10.1561/2200000016]

[3]   E.J. Topol, "High-performance medicine: The convergence of human and artificial intelligence", *Nat. Med.*, vol. 25, no. 1, pp. 44-56, 2019.
[http://dx.doi.org/10.1038/s41591-018-0300-7] [PMID: 30617339]

[4]   J. Camacho, M. Zanoletti-Mannello, Z. Landis-Lewis, S.L. Kane-Gill, and R.D. Boyce, "A conceptual framework to study the implementation of clinical decision support systems (BEAR): literature review and concept mapping", *J. Med. Internet Res.*, vol. 22, no. 8, p. e18388, 2020.
[http://dx.doi.org/10.2196/18388] [PMID: 32759098]

[5]   A. Al Shafei, H. Zareipour, and Y. Cao, *A review of high-performance computing and parallel techniques applied to power systems optimization*. 2022, arXiv.Org, abs/2207.02388.

[6]   S. Maleki Varnosfaderani, and M. Forouzanfar, "The role of AI in hospitals and clinics: Transforming healthcare in the 21st century", *Bioengineering (Basel)*, vol. 11, no. 4, p. 337, 2024.
[http://dx.doi.org/10.3390/bioengineering11040337] [PMID: 38671759]

[7]   P. Hijma, S. Heldens, A. Sclocco, B. van Werkhoven, and H.E. Bal, "Optimization techniques for GPU programming", *ACM Comput. Surv.*, vol. 55, no. 11, pp. 1-239, 2023.
[http://dx.doi.org/10.1145/3570638]

[8]   A.B. Smith, and C.D. Jones, "Optimizing throughput in HPC-based clinical decision support systems", *J. Health Inform.*, vol. 15, no. 2, pp. 345-357, 2010.

[9]   R. Li, J. Liu, G. Zhang, C. Gong, B. Yang, and Y. Liang, "An efficient heterogeneous parallel algorithm of the 3D MOC for multizone heterogeneous systems", *Comput. Phys. Commun.*, vol. 292, p. 108806, 2023.
[http://dx.doi.org/10.1016/j.cpc.2023.108806]

[10]   S. Hermes, T. Riasanow, E.K. Clemons, M. Böhm, and H. Krcmar, "The digital transformation of the healthcare industry: exploring the rise of emerging platform ecosystems and their influence on the role of patients", *Business Research,* vol. 13, no. 3, pp. 1033-1069, 2020.
[http://dx.doi.org/10.1007/s40685-020-00125-x]

[11]   C. Rudin, "Stop explaining black box machine learning models for high stakes decisions and use interpretable models instead", *Nat. Mach. Intell.,* vol. 1, no. 5, pp. 206-215, 2019.
[http://dx.doi.org/10.1038/s42256-019-0048-x] [PMID: 35603010]

[12]   P.S. Pacheco, *Introduction to parallel computing.* 2nd ed. Morgan Kaufmann, 2019.

[13]   Z. Khan, A. Anjum, K. Soomro, and M.A. Tahir, "Towards cloud based big data analytics for smart future cities", *J. Cloud Comput. (Heidelb.),* vol. 4, no. 1, p. 2, 2015.
[http://dx.doi.org/10.1186/s13677-015-0026-8]

[14]   P. Gupta, A. Sharma, and R. Jindal, "Scalable machine-learning algorithms for big data analytics: a comprehensive review", *Wiley Interdiscip. Rev. Data Min. Knowl. Discov.,* vol. 6, no. 6, pp. 194-214, 2016.
[http://dx.doi.org/10.1002/widm.1194]

[15]   E.E. Schadt, M.D. Linderman, J. Sorenson, L. Lee, and G.P. Nolan, "Computational solutions to large-scale data management and analysis", *Nat. Rev. Genet.,* vol. 11, no. 9, pp. 647-657, 2010.
[http://dx.doi.org/10.1038/nrg2857] [PMID: 20717155]

[16]   M. Wernick, Y. Yang, J. Brankov, G. Yourganov, and S. Strother, "Machine learning in medical imaging", *IEEE Signal Process. Mag.,* vol. 27, no. 4, pp. 25-38, 2010.
[http://dx.doi.org/10.1109/MSP.2010.936730] [PMID: 25382956]

[17]   S. Reddy, J. Fox, and M.P. Purohit, "Artificial intelligence-enabled healthcare delivery", *J. R. Soc. Med.,* vol. 112, no. 1, pp. 22-28, 2019.
[http://dx.doi.org/10.1177/0141076818815510] [PMID: 30507284]

[18]   M. G. Kahn, T. J. Callahan, J. Barnard, A. E. Bauck, J. Brown, B. N. Davidson, H. Estiri, C. Goerg, E. Holve, S. G. Johnson, S. T. Liaw, M. Hamilton-Lopez, D. Meeker, T. C. Ong, P. Ryan, N. Shang, N. G. Weiskopf, C. Weng, M. N. Zozus, and L. Schilling, "A harmonized data quality assessment terminology and framework for the secondary use of electronic health record data", *EGEMS (Wash. DC),* vol. 7, no. 1, pp. 1-12, 2019.
[http://dx.doi.org/]

[19]   R. Lenz, and M. Reichert, "IT support for healthcare processes – premises, challenges, perspectives", *Data Knowl. Eng.,* vol. 61, no. 1, pp. 39-58, 2007.
[http://dx.doi.org/10.1016/j.datak.2006.04.007]

[20]   *Exascale computing project: Project plan,* 2021.https://exascaleproject.org/wp-content/uploads/2021/05/ECP_ProjectPlan_2021.pdf

[21]   M. D. Wilkinson, M. Dumontier, I. J. J. Aalbersberg, G. Appleton, M. Axton, A. Baak, N. Blomberg, J.-W. Boiten, L. Bonino da Silva Santos, P. E. Bourne, J. Bouwman, A. J. Brookes, T. Clark, M. Crosas, I. Dillo, O. Dumon, S. Edmunds, C. T. Evelo, R. Finkers, A. Gonzalez-Beltran, A. J. G. Gray, P. Groth, C. Goble, J. S. Grethe, J. Heringa, P. A. C. 't Hoen, R. Hooft, T. Kuhn, R. Kok, J. Kok, S. J. Lusher, M. E. Martone, A. Mons, A. L. Packer, B. Persson, P. Rocca-Serra, M. Roos, R. van Schaik, S.-A. Sansone, E. Schultes, T. Sengstag, T. Slater, G. Strawn, M. A. Swertz, M. Thompson, J. van der Lei, E. van Mulligen, J. Velterop, A. Waagmeester, P. Wittenburg, K. Wolstencroft, J. Zhao, and B. Mons. "The FAIR guiding principles for scientific data management and stewardship," *Scientific Data,* vol. 3, no. 160018, Mar. 2016.
[http://dx.doi.org/10.1038/sdata.2016.18] [PMID: 26978244]

[22]   Z. Obermeyer, and E.J. Emanuel, "Predicting the future — big data, machine learning, and clinical medicine", *N. Engl. J. Med.,* vol. 375, no. 13, pp. 1216-1219, 2016.
[http://dx.doi.org/10.1056/NEJMp1606181] [PMID: 27682033]

[23]   The Atomwise AIMS Program. AI is a viable alternative to high throughput screening: a 318-target study, *Sci. Rep.,* vol. 14, no. 1, p. 7526, 2024.
[http://dx.doi.org/10.1038/s41598-024-54655-z] [PMID: 38565852]

[24]   S. Khanra, A. Dhir, A.K.M.N. Islam, and M. Mäntymäki, "Big data analytics in healthcare: a systematic literature review", *Enterprise Inf. Syst.,* vol. 14, no. 7, pp. 878-912, 2020.
[http://dx.doi.org/10.1080/17517575.2020.1812005]

[25]   I.E. Suleimenov, Y.S. Vitulyova, A.S. Bakirov, and O.A. Gabrielyan, "Artificial intelligence: What is it?", *Proc 2020 6ᵗʰ Int Conf Comput Technol Appl,* pp. 22-5, 2020.
[http://dx.doi.org/10.1145/3397125.3397141]

[26]   T. Li, N. Ferraro, B. J. Strober, F. Aguet, S. Kasela, M. Arvanitis, B. Ni, L. Wiel, E. Hershberg, K. Ardlie, D. E. Arking, R. L. Beer, J. Brody, T. W. Blackwell, C. Clish, S. Gabriel, R. Gerszten, X. Guo, N. Gupta, W. C. Johnson, T. Lappalainen, H. J. Lin, Y. Liu, D. A. Nickerson, G. Papanicolaou, J. K. Pritchard, P. Qasba, A. Shojaie, J. Smith, N. Sotoodehnia, K. D. Taylor, R. P. Tracy, D. Van Den Berg, M. T. Wheeler, S. S. Rich, J. I. Rotter, A. Battle, and S. B. Montgomery, "The functional impact of rare variation across the regulatory cascade", *Cell Genom.,* vol. 3, no. 10, p. 100401, 2023.
[http://dx.doi.org/10.1016/j.xgen.2023.100401]

[27]   K. E. Brown and J. K. Kelly, "Genome-wide association mapping of transcriptome variation in Mimulus guttatus indicates differing patterns of selection on cis- versus trans-acting mutations", *Genetics,* vol. 220, no. 1, p. iyab189, 2022.
[http://dx.doi.org/10.1093/genetics/iyab189]

[28]   M.R. Avendi, A. Kheradvar, and H. Jafarkhani, "A combined deep-learning and deformable-model approach to fully automatic segmentation of the left ventricle in cardiac MRI", *Med. Image Anal.,* vol. 30, pp. 108-119, 2016.
[http://dx.doi.org/10.1016/j.media.2015.12.003]

[29]   Y. Jing, Y. Bian, Z. Hu, L. Wang, and X.Q.S. Xie, "Deep learning for drug design: An artificial intelligence paradigm for drug discovery in the big data era", *AAPS J.,* vol. 20, no. 3, p. 58, 2018.
[http://dx.doi.org/10.1208/s12248-018-0210-0] [PMID: 29603063]

[30]   Y. Wang, L. A. Kung, and T. A. Byrd, "Big data analytics: Understanding its capabilities and potential benefits for healthcare organizations", *Technol. Forecast. Soc. Change,* vol. 126, pp. 3-13, 2018.
[http://dx.doi.org/10.1016/j.techfore.2015.12.019]

[31]   Y. Xu, "Practical implications of hybrid algorithmic optimization in HPC-based CDSS", *Health Informatics J.,* vol. 25, no. 4, pp. 345-358, 2018.
[http://dx.doi.org/10.1177/1460458218761588]

[32]   Y. Xu, "Practical Implications of Hybrid Algorithmic Optimization in HPC-based CDSS", *Health Informatics J.,* vol. 25, no. 4, pp. 345-358, 2018.

[33]   J. Chen, C. Lu, H. Huang, D. Zhu, Q. Yang, J. Liu, Y. Huang, A. Deng, and X. Han, "Cognitive computing-based CDSS in medical practice", *Health Data Sci.,* vol. 2021, p. 9819851, 2021.
[http://dx.doi.org/10.34133/2021/9819851] [PMID: 38487503]

[34]   A.M. Antoniadi, Y. Du, Y. Guendouz, L. Wei, C. Mazo, B.A. Becker, and C. Mooney, "Current challenges and future opportunities for XAI in machine learning-based clinical decision support systems: A systematic review", *Appl. Sci. (Basel),* vol. 11, no. 11, p. 5088, 2021.
[http://dx.doi.org/10.3390/app11115088]

[35]   M. Pandey, M. Fernandez, F. Gentile, O. Isayev, A. Tropsha, A.C. Stern, and A. Cherkasov, "The transformational role of GPU computing and deep learning in drug discovery", *Nat. Mach. Intell.,* vol. 4, no. 3, pp. 211-221, 2022.
[http://dx.doi.org/10.1038/s42256-022-00463-x]

[36]   Z. Shireen, H. Weeratunge, A. Menzel, A.W. Phillips, R.G. Larson, K. Smith-Miles, and E. Hajizadeh, "A machine learning enabled hybrid optimization framework for efficient coarse-graining of a model

polymer", *npj Comput. Mater.,* vol. 8, no. 224, 2022.
[http://dx.doi.org/10.1038/s41524-022-00914-4]

[37]   Z. Zhang, X. Lin, and S. Wu, "A hybrid algorithm for clinical decision support in precision medicine based on machine learning", *BMC Bioinformatics,* vol. 24, no. 1, p. 3, 2023.
[http://dx.doi.org/10.1186/s12859-022-05116-9] [PMID: 36597033]

[38]   L. Kumar Singh, M. Khanna, and R. singh, "A novel enhanced hybrid clinical decision support system for accurate breast cancer prediction", *Measurement,* vol. 221, p. 113525, 2023.
[http://dx.doi.org/10.1016/j.measurement.2023.113525]

[39]   A. Javeed, L. Ali, A. M. Seid, A. Ali, D. Khan, Y. Imrana, and M. Z. Asghar, "A clinical decision support system (CDSS) for unbiased prediction of caesarean section based on features extraction and optimized classification", *Intell. Neurosci.,* vol. 2022, no. 1901735, 2022.
[http://dx.doi.org/10.1155/2022/1901735]

[40]   X. Wang, and Y. Zhang, "Challenges and solutions in algorithmic optimization for HPC architectures", In: *Proc. Int. Conf. High-Performance Comput. (HPC),* 2017, pp. 78-85.
[http://dx.doi.org/10.1145/3126908]

[41]   A. Sateesan, and S. Sinha, "S. K. G., and A. P. Vinod, "A survey of algorithmic and hardware optimization techniques for vision convolutional neural networks on FPGAs,"", *Neural Process. Lett.,* vol. 53, no. 3, pp. 2331-2377, 2021.
[http://dx.doi.org/10.1007/s11063-021-10458-1]

[42]   A. El Bouazzaoui, A. Hadjoudja, O. Mouhib, and N. Cherkaoui, "FPGA-based ML adaptive accelerator: A partial reconfiguration approach for optimized ML accelerator utilization", *Array (N. Y.),* vol. 21, p. 100337, 2024.
[http://dx.doi.org/10.1016/j.array.2024.100337]

[43]   N.H. Noordin, P.S. Eu, and Z. Ibrahim, ""FPGA implementation of metaheuristic optimization algorithm," e-Prime - Adv. Electr. Eng", *e-Prime - Advances in Electrical Engineering, Electronics and Energy,* vol. 6, p. 100377, 2023.
[http://dx.doi.org/10.1016/j.prime.2023.100377]

[44]   M.R. Ezilarasan, J. Britto Pari, and M.F. Leung, "High performance FPGA implementation of single mac adaptive filter for independent component analysis", *J. Circuits Syst. Comput.,* vol. 32, no. 17, p. 2350294, 2023.
[http://dx.doi.org/10.1142/S0218126623502948]

[45]   M. Siddique, and M. Ashour, "Parallel computing with GPU: An accelerator for data-centric high performance computing", In: *Proc. IEEE Int. Conf. Ind. Electron. Syst. Transp. Robot. (ICIESTR),* 2024, pp. 1-6.
[http://dx.doi.org/10.1109/ICIESTR60916.2024.10798202]

[46]   Y. Zhang, Y. Liu, P. Jiao, Y. Zhou, and T. Wei, "Automatic multi-parameter performance modeling of HPC applications on a new Sunway supercomputer", *IEEE Trans. Parallel Distrib. Syst.,* vol. 34, no. 11, pp. 2965-2977, 2023.
[http://dx.doi.org/10.1109/TPDS.2023.3317296]

[47]   R. Sarma, E. Inanc, M. Aach, and A. Lintermann, "Parallel and scalable AI in HPC systems for CFD applications and beyond", *Frontiers in High Performance Computing,* vol. 2, p. 1444337, 2024.
[http://dx.doi.org/10.3389/fhpcp.2024.1444337]

[48]   M. Lieber, and W.E. Nagel, "Highly scalable SFC-based dynamic load balancing and its application to atmospheric modeling", *Future Gener. Comput. Syst.,* vol. 82, pp. 575-590, 2018.
[http://dx.doi.org/10.1016/j.future.2017.04.042]

[49]   R.T. Potla, "Scalable machine learning algorithms for big data analytics: Challenges and opportunities", *J. Artif. Intell. Res.,* vol. 2, no. 2, pp. 124-141, 2022.

[50]   J. Dean, and S. Ghemawat, "MapReduce", *Commun. ACM,* vol. 51, no. 1, pp. 107-113, 2008.

[http://dx.doi.org/10.1145/1327452.1327492]

[51]   P. Gupta, A. Sharma, and R. Jindal, "Scalable machine-learning algorithms for big data analytics: a comprehensive review", *Wiley Interdiscip. Rev. Data Min. Knowl. Discov.,* vol. 6, no. 6, pp. 194-214, 2016.
[http://dx.doi.org/10.1002/widm.1194]

[52]   M. Zaharia, M. Chowdhury, T. Das, A. Dave, J. Ma, M. McCauley, M.J. Franklin, S. Shenker, and I. Stoica, "Resilient distributed datasets: A fault-tolerant abstraction for in-memory cluster computing", In: *Proceedings of the 9ᵗʰ USENIX conference on Networked Systems Design and Implementation (NSDI'12). USENIX Association, USA,* 2012, p. 2.

[53]   J. Lin and C. Dyer, "Data-intensive text processing with MapReduce", *Synth. Lect. Hum. Lang. Technol,* vol. 10, no. 1, pp. 1-177, 2017.
[http://dx.doi.org/10.2200/S00774ED1V01Y201512HLT029]

[54]   G. Malewicz, M.H. Austern, A.J. Bik, J.C. Dehnert, I. Horn, N. Leiser, and G. Czajkowski, "Pregel: A system for large-scale graph processing", *Proc. ACM SIGMOD Int. Conf. Manage. Data,* pp. 135-146, 2010.
[http://dx.doi.org/10.1145/1807167.1807184]

[55]   G. Malewicz, M.H. Austern, A.J. Bik, J.C. Dehnert, I. Horn, N. Leiser, and G. Czajkowski, "Pregel: A system for large-scale graph processing", In: *Proc. ACM SIGMOD Int. Conf. Manage. Data*, 2010, pp. 135-146.
[http://dx.doi.org/10.1145/1807167.1807184]

[56]   Shiv Verma, Luke M. Leslie, Yosub Shin, and Indranil Gupta, "An experimental comparison of partitioning strategies in distributed graph processing", *Proc. VLDB Endow,* vol. 10, no. 5, pp. 493-504, 2017.
[http://dx.doi.org/10.14778/3055540.3055543]

[57]   M. Chen, S. Mao, and Y. Liu, "Big data: A survey", *Mob. Netw. Appl.,* vol. 19, no. 2, pp. 171-209, 2014.
[http://dx.doi.org/10.1007/s11036-013-0489-0]

[58]   M. Zaharia, A. Konwinski, M.J. Franklin, S. Shenker, and I. Stoica, "Resilient Distributed Datasets: A Fault-Tolerant Abstraction for In-Memory Cluster Computing", In: *Proc. 9ᵗʰ USENIX Conf. Netw. Syst. Des. Implement.* vol. 2. , 2012no. 2, .
[http://dx.doi.org/10.1145/2208914.2208949]

[59]   T. Benacchio, L. Bonaventura, M. Altenbernd, C.D. Cantwell, P.D. Düben, M. Gillard, L. Giraud, D. Göddeke, E. Raffin, K. Teranishi, and N. Wedi, "Resilience and fault tolerance in high-performance computing for numerical weather and climate prediction", *Int. J. High Perform. Comput. Appl.,* vol. 35, no. 4, pp. 285-311, 2021.
[http://dx.doi.org/10.1177/1094342021990433]

[60]   S. U. Swetha and P. M. K. Prakruthi, "A big data implementation on grid computing", *Int. J. Eng. Res. Technol. (IJERT)*, vol. 2, no. 13, NCRTS, 2014.

[61]   J. Lee, K. Park, H. Kim, and J. Choi, "The importance of data accuracy in predictive modeling", *J. Healthc. Inform. Res.,* vol. 1, no. 1, pp. 25-36, 2017.
[http://dx.doi.org/10.1016/j.jhir.2016.12.003]

[62]   T. Hey, and S. Tansley, *The Fourth Paradigm: Data-Intensive Scientific Discovery.* Microsoft Research: Redmond, WA, USA, 2009.

[63]   S. Shankar, and A.K. Sharma, "A comparative performance analysis of cloud, cluster and grid computing over network", *Int. J. Eng. Res. Technol. (IJERT). ICADEMS,* vol. 5, no. 03, p. 2017, 2017.
[http://dx.doi.org/10.17577/IJERTCONV5IS03046]

[64]   S. Wang, H. Zheng, X. Wen, and S. Fu, "Distributed high-performance computing methods for accelerating deep learning training", *J. Knowl. Learn. Sci. Technol.,* vol. 3, no. 3, pp. 108-126, 2024.

[http://dx.doi.org/10.60087/jklst.v3.n3.p108-126]

[65]   T. Subramanian, "Integrating machine learning in clinical decision support systems", *Health Informatics: An Int. J. (HIIJ),* vol. 13, no. 1, pp. 1-11, 2024.
[http://dx.doi.org/]

[66]   P. Cheng, Y. Lu, Y. Du, and Z. Chen, "Tiered data management system: Accelerating data processing on HPC systems", *Future Gener. Comput. Syst.,* vol. 101, pp. 894-908, 2019.
[http://dx.doi.org/10.1016/j.future.2019.07.046]

[67]   B.A. Van Dort, W.Y. Zheng, V. Sundar, and M.T. Baysari, "Optimizing clinical decision support alerts in electronic medical records: a systematic review of reported strategies adopted by hospitals", *J. Am. Med. Inform. Assoc.,* vol. 28, no. 1, pp. 177-183, 2021.
[http://dx.doi.org/10.1093/jamia/ocaa279] [PMID: 33186438]

[68]   P. Balaprakash and J. Dongarra, "Autotuning in high-performance computing applications", *Proc. IEEE,* vol. 106, no. 7, pp. 1234-1245, 2018.
[http://dx.doi.org/10.1109/JPROC.2018.2841200]

[69]   J. Dean, and S. Ghemawat, "MapReduce: Simplified data processing on large clusters", *Commun. ACM,* vol. 51, no. 1, pp. 107-113, 2008.
[http://dx.doi.org/10.1145/1327452.1327492]

[70]   P. Hijma, S. Heldens, A. Sclocco, B. van Werkhoven, and H.E. Bal, "Optimization techniques for GPU programming", *ACM Comput. Surv.,* vol. 55, no. 11, p. 81, 2023.
[http://dx.doi.org/10.1145/3570638]

[71]   J. Dümmler, T. Rauber, and G. Rünger, "Scalable computing with parallel tasks", In: *Proc. 2nd Workshop Many-Task Comput. Grids Supercomputers (MTAGS '09)* vol. 9. New York, NY, USA: Association for Computing Machinery, 2009, pp. 1-10.
[http://dx.doi.org/10.1145/1646468.1646477]

[72]   S.P. Heilbroner, and R. Miotto, "Deep learning in medicine", *Clin. J. Am. Soc. Nephrol.,* vol. 18, no. 3, pp. 397-399, 2023.
[http://dx.doi.org/10.2215/CJN.0000000000000080] [PMID: 36735512]

[73]   C.A. Sinsky, H. Bavafa, R.G. Roberts, and J.W. Beasley, "Standardization *vs* customization: Finding the right balance", *Ann. Fam. Med.,* vol. 19, no. 2, pp. 171-177, 2021.
[http://dx.doi.org/10.1370/afm.2654] [PMID: 33685879]

[74]   E.L. Sevigny, J. Greathouse, and D.N. Medhin, "Health, safety, and socioeconomic impacts of cannabis liberalization laws: An evidence and gap map", *Campbell Syst. Rev.,* vol. 19, no. 4, p. e1362, 2023.
[http://dx.doi.org/10.1002/cl2.1362] [PMID: 37915420]

[75]   European Commission, *Ethics guidelines for trustworthy AI,* 2018.https://ec.europa.eu/digital-strategy

[76]   R. Sedgewick, *Algorithms in C.* Addison-Wesley: Boston, MA, USA, 2001.

CHAPTER 4

Integration of Internet of Medical Things (IoMT) and Artificial Intelligence Applications in Healthcare and Medicine – A Multi-Disciplinary Perspective

R. Raffik[1,*], **L. Feroz Ali**[2], **W. Aagasha Maria**[1], **B. Subashini**[1] and **R. Asvitha**[1]

[1] *Department of Mechatronics Engineering, Kumaraguru College of Technology, Coimbatore, Tamil Nadu, India*

[2] *Department of Mechatronics Engineering, Sri Krishna College of Engineering and Technology, Coimbatore, Tamil Nadu, India*

Abstract: Advanced information technologies drive innovation, with the Internet of Things (IoT) impacting medicine and healthcare sectors. The delivery of patient care and medical procedures has been transformed by the adoption of IoT technologies. It assists in medical records management, aids in understanding disease causes, and early detection of the disease, which in turn give people the confidence to take charge of their well-being. The utilization of wearable health trackers, implantable sensors, and smart medical devices facilitated by IoT has significantly eased real-time monitoring, diagnosis, and treatment optimization, marking a revolutionary shift toward preventive healthcare. Virtual telemedicine technologies like video conferencing and remote health monitoring systems reduce in-person doctor appointments. Telemedicine lowers treatment expenses and saves time for patients and medical professionals alike. Telemedicine devices with remote patient monitoring hardware provide great video quality for real-time predictive healthcare systems. Medical surgical robots execute precise surgical tasks with exceptional accuracy and control. Robotic Surgical System facilitates surgeons to precisely control the surgeries through magnified 3D vision, and wrist-mounted controls, enabling precise incisions.

Minimally Invasive Neurosurgical Intracranial Robot with cutting-edge robotic systems using Shape memory alloy actuators are utilized for brain tumor removal. A cyber-physical system (CPS) is introduced which includes continuous monitoring of vital health parameters (*e.g.*, blood glucose, blood pressure) and triggers automated treatment when critical levels are detected, easing patient care. AI diagnostic tools, using image recognition and natural language processing, improve medical decisions and patient outcomes. This chapter discusses the Internet of Medical Things, Surgical

---

* **Corresponding author R. Raffik:** Department of Mechatronics Engineering, Kumaraguru College of Technology, Coimbatore, Tamil Nadu, India; E-mail: raffik.r.mce@kct.ac.in

Parikshit N. Mahalle, Gitanjali R. Shinde, Namrata N. Wasatkar & Prashant R. Anerao (Eds.)

Robots, Medical Robot Actuation and Control of Sensors and Actuators, Cyber-Physical Health Systems, and Artificial Intelligence applications in healthcare sectors like disease identification and its remedial preventive actions, sensorial fusion techniques in healthcare sectors.

Keywords: Cyber-physical health systems, Healthcare automation, IoMT, Preventive healthcare, Surgical medical robots, Telemedicine.

INTRODUCTION

In a time of exponential technological development, the healthcare industry is at the forefront of change. The combination of the Internet of Medical Things (IoMT), telemedicine, and healthcare automation has ushered in a new era in healthcare delivery and management. At the same time, cyber-physical health systems, medical-surgical robots, and preventive health interventions have become key elements of a paradigm shift towards more efficient, accessible, and patient-centered care. This chapter describes the combination of these disruptive technologies and provides a comprehensive overview of their overall impact on the healthcare industry. From leveraging IoMT for remote patient monitoring to seamless delivery of healthcare services *via* telemedicine platforms, every innovation plays a key role in improving the accessibility and efficiency of healthcare. Additionally, implementing automation in healthcare simplifies administrative tasks and allows healthcare professionals to allocate more resources to provide patient-centered care. This chapter details how cyber-physical healthcare systems and medical-surgical robots are transforming surgical techniques, diagnostics, and treatments. Additionally, the emerging topic of health care is explored and the importance of a proactive approach to preventing disease and promoting well-being is highlighted. This chapter provides a detailed overview of the disruptive innovations transforming the healthcare landscape through in-depth analysis and illustrative case studies. This is a valuable resource for healthcare providers, researchers, policymakers, and technologists, providing insights into how these innovations can be used to improve patient outcomes and reshape the future of healthcare.

EVOLUTION OF IOMT FROM IOT

The development of the Internet of Medical Things (IoMT) from the Internet of Things (IoT) signifies a major advancement in applying connected technology within the healthcare sector. Initially, IoT involved a network of physical objects with embedded sensors, software, and other technologies to gather and exchange data over the Internet. These technologies were first applied in diverse fields such as smart homes, industrial automation, and smart cities, focusing on improving efficiency, automation, and data-driven decision-making [1]. In healthcare, early

IoT applications included wearable devices like fitness trackers and smartwatches that monitored physical activity, heart rate, and sleep patterns. These devices provided essential health data to users and healthcare providers, promoting a proactive approach to health management. Additionally, remote monitoring tools were developed to track vital signs like blood pressure, glucose levels, and oxygen saturation. These tools allowed for continuous monitoring of patients with chronic conditions, minimizing the need for frequent hospital visits and enabling more timely medical interventions [2].

Building on these initial technologies, IoMT emerged as a specialized branch of IoT focused solely on healthcare applications. This transition led to the creation of advanced connected medical devices and applications, including sophisticated wearable health monitors, implantable devices, smart pills, and telemedicine platforms. For example, advanced wearable health monitors can track a variety of health metrics in real-time, such as electrocardiograms (ECG) and blood oxygen levels, allowing for continuous health monitoring and early detection of potential issues. Implantable devices, like pacemakers and insulin pumps, provide critical data and adjust their functions based on real-time monitoring, thereby improving patient outcomes. Smart pills, equipped with ingestible sensors, can track medication adherence and provide data on how drugs interact within the body, enhancing treatment effectiveness. Telemedicine platforms have also become an essential component of IoMT, enabling patients to consult with healthcare providers remotely, thus increasing access to medical care, especially in underserved areas [3].

IoMT not only enhances real-time health monitoring and personalized treatment plans but also significantly improves data management and the accuracy of medical records. This interconnected network of devices facilitates seamless communication among healthcare providers, leading to more coordinated and efficient care delivery. Moreover, IoMT leverages advancements in big data analytics and artificial intelligence (AI) to derive valuable insights from the vast amounts of data generated by connected devices [4]. These technologies support predictive analytics, which can anticipate disease outbreaks, improve the management of chronic illnesses, and optimize resource allocation in healthcare settings.

The integration of AI with IoMT enables advanced data analysis, identifying patterns and trends that may not be immediately evident to human analysts. For example, AI algorithms can predict patient deterioration based on subtle changes in vital signs, prompting early interventions that can prevent complications and hospitalizations. Additionally, AI-powered decision support systems can assist healthcare providers in diagnosing conditions and recommending treatment plans, thereby improving the overall quality of care. The paradigm shift from IoT to IoMT marks a significant step towards a more interconnected, data-driven, and patient-centric healthcare ecosystem. Fig. (**1**) shows the integration of wearable medical sensors with wearable devices and clothing. By utilizing connected technologies, IoMT aims to improve patient outcomes, reduce healthcare costs, and enhance the overall efficiency of healthcare delivery. This evolution highlights the potential of IoMT to transform healthcare, making it more proactive, personalized, and responsive to the needs of patients [5].

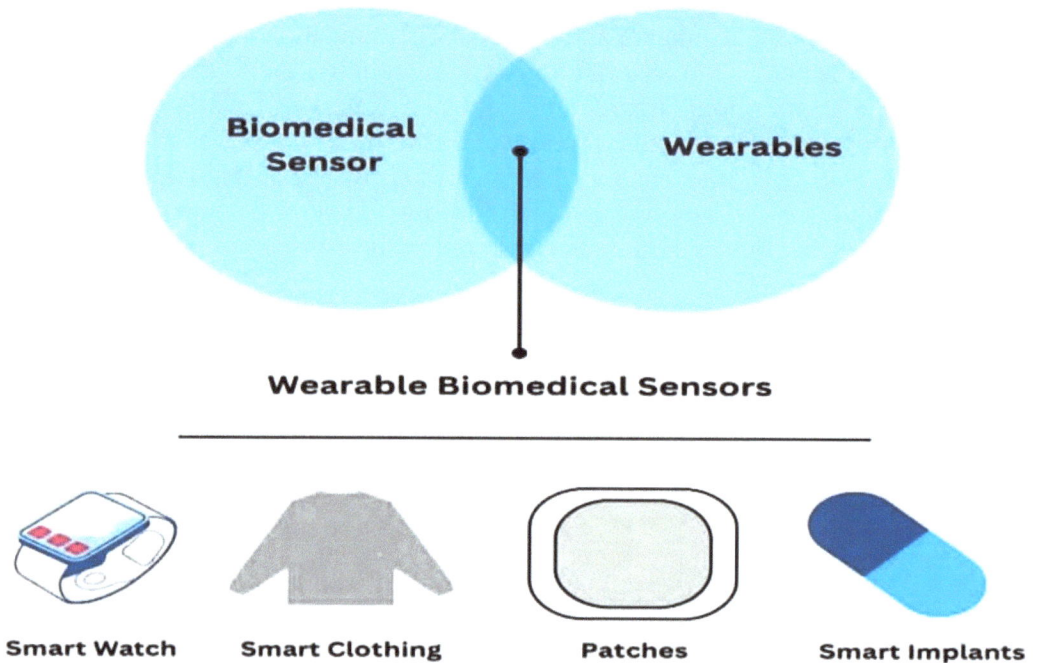

Fig. (1). Wearable biomedical sensors.

One emerging advancement involves the fusion of EHRs (electronic health records) with wearable health technology, offering a unique capability for healthcare systems. While wearable devices hold promise in revolutionizing patient care, obstacles such as worries regarding patient confidentiality, compatibility between systems, and managing the influx of patient data present hurdles to providers considering the adoption of wearables. An optimal wearable device system should be deployable either intermittently or continuously, boasting a prolonged battery lifespan and minimal energy consumption. It must transmit health-related data wirelessly while allowing users unrestricted movement. Calibration should be regularly upheld to mitigate the impact of factors like temperature, humidity, and motion especially due to the textile electrodes used [6]. Despite their essential attributes, wearable technologies are undergoing rapid evolution. These are built using biomedical sensors combined with any wearable accessory. Wearable sensors incorporate three-dimensional (3D) sensors like gyroscopes, accelerometers, and magnetometers.

THE ARCHITECTURE OF WEARABLE HEALTH DEVICES

Primarily constituted of a wearable sensor system, a network transmission system, and an information processing system, structured upon the three-layer architecture of the Internet of Things.

The data collected are transmitted using Bluetooth technology to access the real-time data remotely [7]. Fig. (**2**) shows the architecture of the patient monitoring system using wearable sensors to monitor Heart rate, Air Intake Quality, SpO_2, Respiratory system function, *etc.*, The wearable sensor system collects health data *via* wearable designs. It contains medical sensors such as pulse meters, thermometers, skin response electrodes, accelerometers, and blood pressure sensors, which measure blood pressure, oxygen saturation, blood sugar, pulse rate, heart rate, and body temperature. Data is wirelessly transported to a gateway where it is processed, displayed, and monitored in real time. The network transmission system is used for information transmission. Home gateway and mobile gateway systems are utilized. Mobile gateways, like cell phones, give mobility outside and, home gateways, which are usually computers, are stable for family contexts. The information processing system manages the analysis and storage of data. A database server receives physiological data instantly. Medical facility managers can view the data shown on specialized equipment once it has been processed by pertinent software. When indicators surpass thresholds, alarm values trigger actions. To help avoid sickness, real-time data is evaluated and saved for health reports and recommendations.

Common technologies for wireless communication include Bluetooth, Wi-Fi, and ZigBee. Wi-Fi delivers high transmission rates and ranges, Bluetooth 4.0 saves electricity, and ZigBee is ideal for big sensor networks. Several variables, including cost, topology, transmission rate, range, and power consumption, affect selection [8]. The system's human body sensor network consists of multiple sensor nodes that measure parameters such as blood pressure, blood oxygen saturation, blood sugar, pulse, and heart rate. This setup eliminates the need for extensive networking.

Architecture based on Cloud computing is developing where Body Sensor Networks (BSNs) provide seamless data flow from sensors to cloud servers by combining wearable devices with cloud infrastructure. This ensures secure storage and processing of the collected data. The scalability and computational capability needed to handle the large amount of physiological data produced by BSNs are provided by cloud computing. Furthermore, the application of machine learning techniques and cloud-based analytics algorithms makes it possible to extract valuable insights from the data, promoting personalized interventions, early detection of health issues, and predictive modeling to improve healthcare outcomes [9]. Body Sensor Networks (BSNs) are revolutionizing healthcare delivery by utilizing worn sensors and cloud computing to create collaborative wearable sensor networks that offer personalized treatment plans and ongoing monitoring to improve overall health.

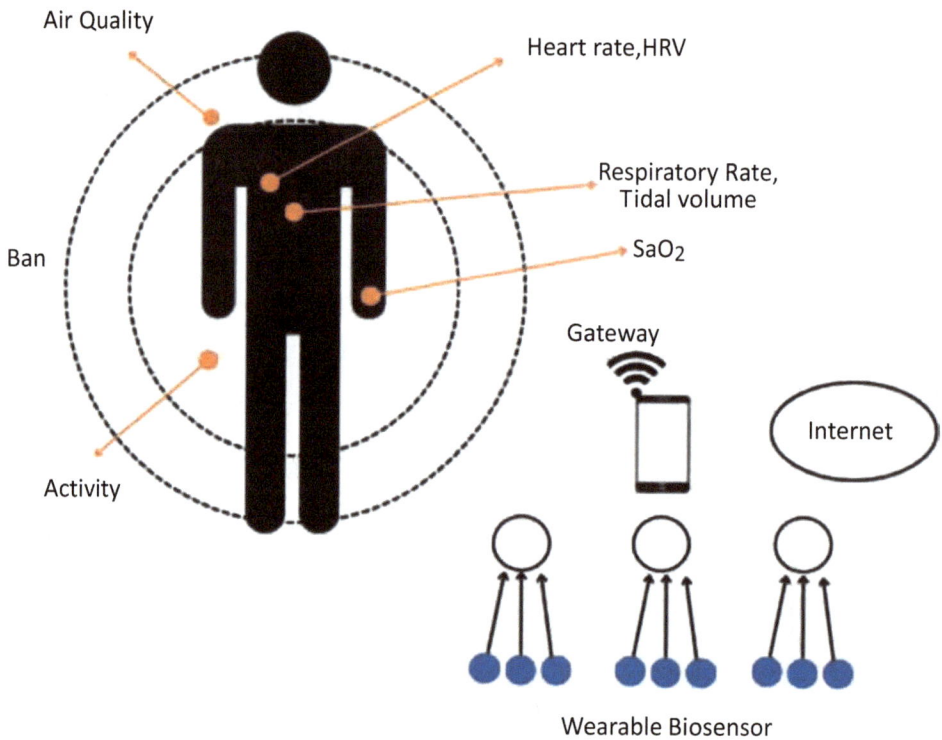

Fig. (2). Architecture of patient monitoring system using wearable sensor.

Exploring Cardio and Respiratory Systems with Wearable Sensor Technology

Sports have long employed heart rate monitoring, usually with the use of chest strap electrodes. On the other hand, photoplethysmography-based wrist-worn devices now offer more convenient continuous monitoring. While each device's accuracy varies, newer models usually include peak detection algorithms, which lead to median error rates of less than 5% when compared to telemetry in healthy subjects. However, because they modify photoplethysmography signals, irregular rhythms like atrial fibrillation can make reliable measurements more difficult to obtain. Complex algorithms have been devised to tackle this problem. Even ECG recording features and a limited amount of rate and rhythm analysis software are available on modern smartwatches. Moreover, the development of wearables that run on their energy using triboelectric nanogenerator technology presents a hope for continuous heart rate monitoring.

The sensors demonstrate a strong correlation with ECG signals and harness biomechanical energy from wrist motion. In managing congestive heart failure (CHF), decreased activity levels serve as predictors for adverse outcomes, including mortality. Wearable actigraphy devices have been validated against patient diaries and physical activity during cardiac rehabilitation. Actigraphy from wrist-based devices was used to classify 50 CHF patients into high and low-physical activity groups, revealing significantly higher rates of hospital admissions in the low-activity group. A systematic review highlighted that reduced physical activity in CHF patients correlated with unfavorable clinical outcomes and mortality.

In the AWAKE-HF study, which compared sacubitril and valsartan with enalapril in heart failure patients, actigraphy was employed to monitor physical activity. Although patient-reported quality of life improved in the sacubitril and valsartan groups, there was no significant difference in physical activity detected between treatment groups. Furthermore, a band electrode integrated into clothing can measure changes in intrathoracic impedance, potentially offering more reliable predictions of hospital admissions compared to changes in weight alone.

Daily ReDS (lung fluid monitoring system) surveillance in 50 CHF patients over 90 days led to significant reductions in admissions, as supported by retrospective studies. Photoplethysmography is employed for oxygen saturation measurement, while fitness gadgets commonly estimate maximal oxygen consumption (VO_2 max) during physical activity. Respiratory rate assessment is facilitated by smart garments equipped with accelerometers, gyroscopes, or magnetometers, which detect chest wall movements, circumference changes, or impedance pneumography. The adoption of wearable devices for evaluating respiratory function has notably increased since the onset of the COVID-19 pandemic [10].

The use of wearable devices in clinical decision-making lacks validation, requiring further research to understand their impact on medical outcomes. Cardiologists need to integrate wearable data with traditional symptom assessment. Advances in machine learning and sensor technology offer the potential for new insights. Rapid development of regulatory frameworks for digital healthcare is crucial, alongside efforts to build public trust [11]. Wearables could help reduce healthcare disparities if made accessible to all, and integration with telemedicine could transform community care and reduce healthcare spending.

Table **1** enlists the various sensor technologies utilized in medical sectors illustrated with some examples. Future advancements in digital technology within this domain will depend on ongoing examination of optimal approaches, areas of concern, and prospective remedies to alleviate current obstacles [12].

Table 1. Sensor technology applied in cardiovascular diseases (wearable devices).

Sensor Technology	Applications	Examples of Wearable Devices
Ballistocardiography Sensor	Utilized for studying cardiac dynamics and cuffless blood pressure monitoring.	Incorporated into wristwatches or wristbands
Electrocardiogram (ECG)	Monitors heart activity, including heart rate and rhythm assessment.	Embedded in various wearables such as wristwatches, shirts, or vests.
Impedance-plethysmography	Measures blood pressure without a cuff and assesses thoracic impedance in heart failure cases.	Integrated into wearable devices like wristwatches, wristbands, shirts, or vests
Photoplethysmography	Monitors heart rate, blood oxygen levels, and cuffless blood pressure.	Incorporated into wearables such as wristwatches, wristbands, or eyeglasses.
Accelerometer	Applied for movement analysis, including step tracking, activity monitoring, and fall detection.	Commonly found in smartphones, wristwatches, wristbands, armbands, or belts.

The use of wearable devices in clinical decision-making lacks validation, requiring further research to understand their impact on medical outcomes. Cardiologists need to integrate wearable data with traditional symptom assessment. Advances in machine learning and sensor technology offer the potential for new insights [13]. Rapid development of regulatory frameworks for digital healthcare is crucial, alongside efforts to build public trust. Wearables could help reduce healthcare disparities if made accessible to all, and integration with telemedicine could transform community care and reduce healthcare spending [14]. Future advancements in digital technology within this domain will depend on ongoing examination of optimal approaches, areas of concern, and prospective remedies to alleviate current obstacles.

ROBOTIC PROCESS AUTOMATION (RPA) IN HEALTHCARE OPERATIONS

Robotic Process Automation (RPA) leverages AI and ML to automate repetitive tasks across various sectors, optimizing processes and managing data efficiently. Platforms like UiPath and Automation Anywhere enable seamless implementation of automation solutions. While RPA offers benefits like enhanced accuracy and cost reduction, it also poses challenges such as job displacement and security concerns. ML techniques drive intelligent decision-making within RPA processes, shaping the future of work across industries [15]. RPA transforms healthcare by automating tasks like appointment scheduling and claims processing, boosting efficiency and patient satisfaction. It streamlines operations, cuts errors, and ensures compliance, while also enhancing revenue management and patient registration. Overall, RPA is a game-changer in healthcare operations, delivering benefits for providers and patients alike.

Robotic Surgical System

Medical surgical robots have the potential to enhance surgical results and lower problems, and they are quickly changing the medical sector. These robots enable more precise and successful surgeries because of their special blend of control and precision. The development of robotic platforms and tools specifically suited for general surgical operations is receiving more interest and funding because of the significant benefits that these advancements have brought to both patients and surgeons [16]. Without directly taking from the original text, this study attempts to explore the present and future directions in surgical robotic technologies within a changing research environment. The field of surgical robotics has advanced significantly in recent years, enabling surgeons to perform intricate procedures with greater precision, accuracy, and safety. Computer-controlled surgical robots can assist in various surgeries, including minimally invasive procedures, resulting in fewer scars, shorter hospital stays, and faster recovery times compared to traditional methods [17]. They also provide surgeons with increased dexterity and visual capabilities, enabling them to perform previously risky or difficult treatments. As technology continues to develop, surgical robots are expected to play an even more significant role in medicine, transforming surgical practices and improving patient outcomes. This review will assess the current state of surgical robots, available tools, their use in various specialties, and their impact on patient outcomes [18].

Software Used in Surgical Robots

The integration of diverse sensors, navigation tools, and devices with surgical robots signifies a leap forward toward the evolution of smarter surgical tools.

However, the specialized software underpinning these advancements is frequently proprietary, limiting broader access and innovation. Recognizing this barrier, the Open Core Control software emerged as a solution, designed to facilitate the use of different surgical robots. It manages this by modularizing the elements that rely on hardware, thus broadening compatibility across a range of robotic systems [19].

Additionally, the ability of medical devices to connect and communicate through networks is essential for enhancing their cooperative capabilities. The Interface class leverages Open IGT Link for seamless interaction with other medical devices, guaranteeing consistent performance when exchanging data asynchronously across the network. A pivotal strategy for assisting operators in achieving meticulous control over master-slave robotic systems is the implementation of the virtual fixture method [20]. Demonstrating the collaborative potential between a surgical robot and a navigation system, a virtual fixture add-on was incorporated into the setup. This feature allows operators to designate a specific zone within the navigation system that can be communicated to the robot, facilitating precise surgical maneuvers [21]. During surgery, the surgical console generates resistance if the operator attempts to deviate from the predefined accessible area, ensuring safety and precision. In the rapidly advancing field of medical technology, surgical robots have become key tools, enhancing the skillset of human surgeons with unprecedented levels of accuracy and efficiency. The heart of these transformative devices is a suite of complex software, each part meticulously crafted to play an essential role in the surgical procedure [22].

At the forefront of this technology is the control software, an engineering masterpiece that converts the surgeon's inputs into precise robotic arm and tool movements. This software faithfully carries out the surgeon's directions with remarkable precision, allowing for surgical interventions that exceed the natural capabilities of the human hand. Supporting the control software is the imaging and visualization toolkit, which delivers live, three-dimensional images of the surgery area [23].

This toolkit merges flawlessly with various imaging techniques like MRI and CT scans, providing a holistic view that informs the surgeon's decisions and actions with exceptional clarity. Equally crucial is the simulation and training software, which creates a safe space for surgeons to improve their skills [24]. This virtual practice environment lets surgeons perfect their techniques, investigate new surgical methods, and prepare for intricate surgeries with a degree of realism and detail unachievable with traditional training approaches. Moreover, the navigation and mapping software acts as a guiding light through the human body's complex structures. It uses precise patient data to map out the surgical field, directing instruments with an accuracy that ensures patient safety and bolsters surgical success [25].

Collectively, these software elements are the foundation of surgical robotics, enabling surgeons to conduct minimally invasive procedures with a level of control, adaptability, and precision previously considered science fiction. Table **2** lists the applications of surgical robots and their utilization approaches in the medical sector. As these technologies progress, they promise to revolutionize surgical practices, shorten patient recovery periods, and enhance treatment outcomes across the medical spectrum [26].

Table 2. Overview of studies on surgical robots.

Applications	Approaches
Minimally invasive surgical approach (prostatectomies, cardiac valve repair)	The da Vinci surgical system. Surgical robotics
Cardiac arrhythmias [27]	Steerable catheter
Angiography	Magellan Robotic system
Reduce fatigue and muscle activation	Surgical exoskeleton
Patient-size-adjustable frame	HAL exoskeleton
Remote controlled telemanipulation	Teleoperated surgical robots

Da Vinci R Surgical System

The da Vinci architecture is built around a master-slave configuration that encompasses three main components: the surgeon's console (master), the vision cart (central control unit), and the patient cart (slave unit). Within this surgical framework, the surgeon controls the robot remotely from a console located outside the surgery room. Using hand controls and foot pedals at the console, the surgeon directs the robot's movements. A stereo viewer provides the surgeon with a 3D.

Teleoperated Surgical Robots

The challenge of control involves determining the optimal utilization of a surgical robot's multiple axes of motion to execute a given task within the confines of spatial constraints, the robot's design, and the specific requirements of the task. Due to the variety of operational contexts, modern robotic systems are generally classified into four main types. These categories include the use of force and motion for rehabilitation, automated execution of movements along a predetermined path based on pre-surgical planning, the stationary positioning of instruments and tools, and teleoperated remote control [28]. Among the various systems reviewed, which employ commercially available industrial robots for Task-Oriented Surgery (TOS) and their industrial counterparts, it was noted that the majority rely on articulated robotic arms with a lower carrying capacity [29]. Of these, only three systems boasted seven axes of freedom, while the remainder were limited to six. Among all the evaluated systems, only the MIRO robot is specifically designed for use in medical settings.

ROLE OF NANOTECHNOLOGY AND IOMT IN HEALTHCARE

Understanding the role of nanotechnology is crucial, especially as achieving the desired levels of miniaturization and sensitivity requires an advanced understanding that AI can provide. Integrating nanotechnology with AI systems, further enhanced by the Internet of Medical Things (IoMT) technology, is critical for pioneering healthcare innovations in areas such as nanomedicine and nanorobotics. The fusion of AI with IoMT devices, especially in critical healthcare sectors like cardiac care, surgical interventions, diabetes management, and oncology tracking, is vital. These devices, equipped with nano-modified sensors, continuously track vital health metrics [30]. The emergence of 2D functional materials, including graphene, borophene, and MXenes, has led to the creation of cutting-edge biosensors offering enhanced spatial and temporal resolution. For instance, MXene-based E-Skin sensors that capitalize on pressure sensitivity are adept at tracking human motion. When paired with wireless technology, these sensors are useful in a range of scenarios, from monitoring health treatments to tracking movement [31]. Carbon nanomaterials, such as carbon nanotubes and graphene, are prized for their conductive properties and compatibility with biological systems, playing a pivotal role in the development of wearable screens that can be integrated smoothly onto skin-like surfaces [32]. Furthermore, sensors that mimic human sensory mechanisms, crafted from graphene derivatives, fulfill diverse roles from touch sensitivity to replicating the sense of smell. The melding of flexible printed bio-signal tracking devices with AI enables the wireless, real-time tracking of health metrics, while textile-based sensor [33] technologies facilitate the continuous observation of vital signs, as demonstrated by fabric-based triboelectric sensors that non-invasively monitor blood pressure. These nano-technology-driven sensing methods, augmented by AI's predictive power on IoT frameworks, hold promise for the early detection and management of chronic conditions.

CYBER-PHYSICAL HEALTH SYSTEMS

Research into cyber-physical systems (CPS) in healthcare is still in its infancy. CPS integrates active user inputs such as intelligent feedback systems and digital health records as well as passive inputs such as biosensors and smart health devices, enabling efficient data collection for informed decision-making [34]. This combination of data collection and decision systems is still largely unexplored in healthcare applications and is therefore the focus of research interest. Opportunities for the use of CPS in healthcare include implementing coordinated interaction of autonomous and adaptive devices, developing novel approaches to managing medical systems through computing and control, leveraging miniaturized implantable smart devices and body area networks, and exploring programmable materials and advanced manufacturing techniques.

Architecture

The design of cyber-physical systems (CPS) in healthcare is of considerable importance in ensuring system quality and performance. The cyber-physical health system integrates physical sensors to capture real-time health data, which is then processed and transferred using communication infrastructure. Advanced analytics and machine learning extract insights for decision assistance and provide personalized suggestions *via* user interfaces. Security measures protect sensitive health information, and connection with healthcare systems promotes collaboration. A feedback loop ensures ongoing performance improvement. Fig. (**3**) shows the architecture of Cyber-Physical Health Systems with Sensing, Actuation, Data Management, Service Aware, and Application modules. It illustrates the integration of the cyber world with the physical world.

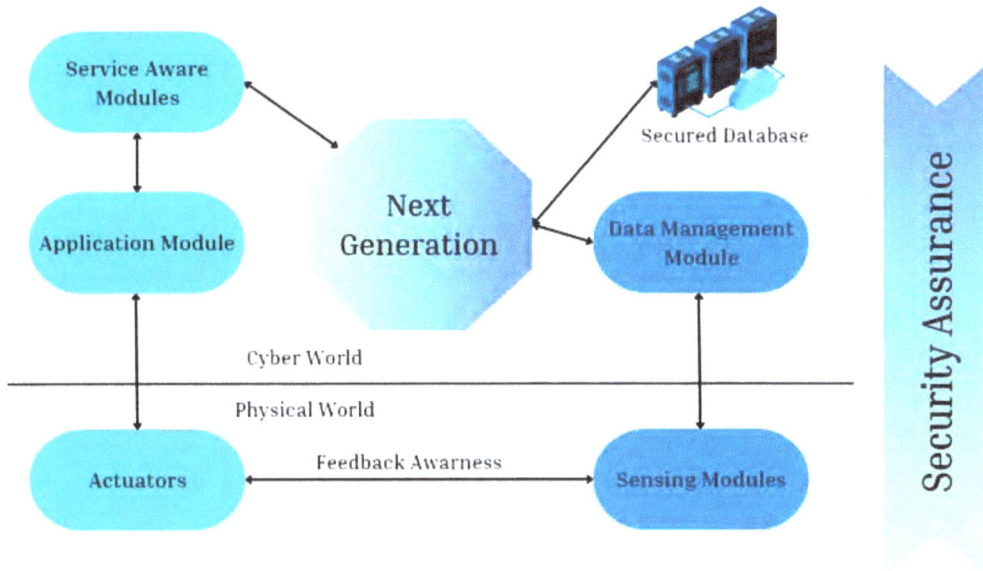

Fig. (3). Architecture of cyber-physical health systems.

At its foundation lies a network of physical sensors and devices strategically placed to collect real-time health data from various origins, including wearable gadgets, medical instruments, and monitoring tools. These sensors continuously gather information on vital signs, activity levels, physiological markers, and other health-related metrics. Once collected, this raw data is transmitted *via* a sturdy communication framework, often utilizing wireless connections or IoT protocols, to designated systems for storage and management. The data then undergoes processing, consolidation, and examination using state-of-the-art algorithms and machine learning techniques. These procedures unveil valuable insights, detect patterns, and unveil correlations within the data, ultimately providing actionable knowledge to guide healthcare decision-making. These insights are translated into practical intelligence and tailored recommendations through user-friendly interfaces, empowering both healthcare providers and patients to make informed decisions regarding health and wellness. This might include suggesting lifestyle changes, adjustments to medication plans, or proactive measures to prevent potential health complications.

Given the sensitive nature of health data, stringent security measures are crucial to uphold its confidentiality, integrity, and accessibility. Encryption, access controls, authentication mechanisms, and data anonymization strategies are utilized to mitigate risks and adhere to privacy regulations such as HIPAA and GDPR. Furthermore, seamless integration with existing healthcare systems is imperative to facilitate collaboration among various stakeholders, including healthcare professionals, researchers, caregivers, and individuals receiving care. Interoperability standards, APIs, and data exchange protocols ensure smooth information flow across diverse platforms and systems, fostering a comprehensive approach to healthcare provision. Additionally, the system incorporates a feedback loop mechanism to continuously monitor performance, gather input, and iteratively improve system capabilities over time. This involves soliciting user feedback, monitoring system performance metrics, assessing outcomes, and applying acquired insights to enhance functionality, usability, and effectiveness.

Advantages and Disadvantages of AI

Fig. (**4**) depicts the various advantages and disadvantages of AI in the Medicinal sector.

- **Access to Real-time Data:** AI enables rapid access to crucial medical information, fostering accurate clinical decision-making and improving doctor-patient relationships.

- **Automation of Tasks:** AI simplifies scheduling, data interpretation, and record-keeping, enhancing efficiency and reducing the workload on healthcare providers.
- **Time and Resource Savings:** AI-powered automation streamlines administrative tasks, potentially saving over $200 billion annually in healthcare costs.
- **Improving Research:** AI contributes to medical research by analyzing vast data, offering valuable expertise in studying and treating dangerous diseases.
- **Reliance on Human Oversight:** Effective AI use requires human input and judgment, necessitating collaboration between medical professionals and technical specialists to enhance AI understandability through expertise [35].
- **Potential Bias:** Limited data on certain diseases, demographics, or environmental variables may lead to errors in AI medical applications, especially in critical areas like drug prescribing [36].
- **Vulnerabilities:** AI-based cyberattacks may become more sophisticated, posing challenges in prediction and prevention as attackers leverage AI to enhance their strategies [37].

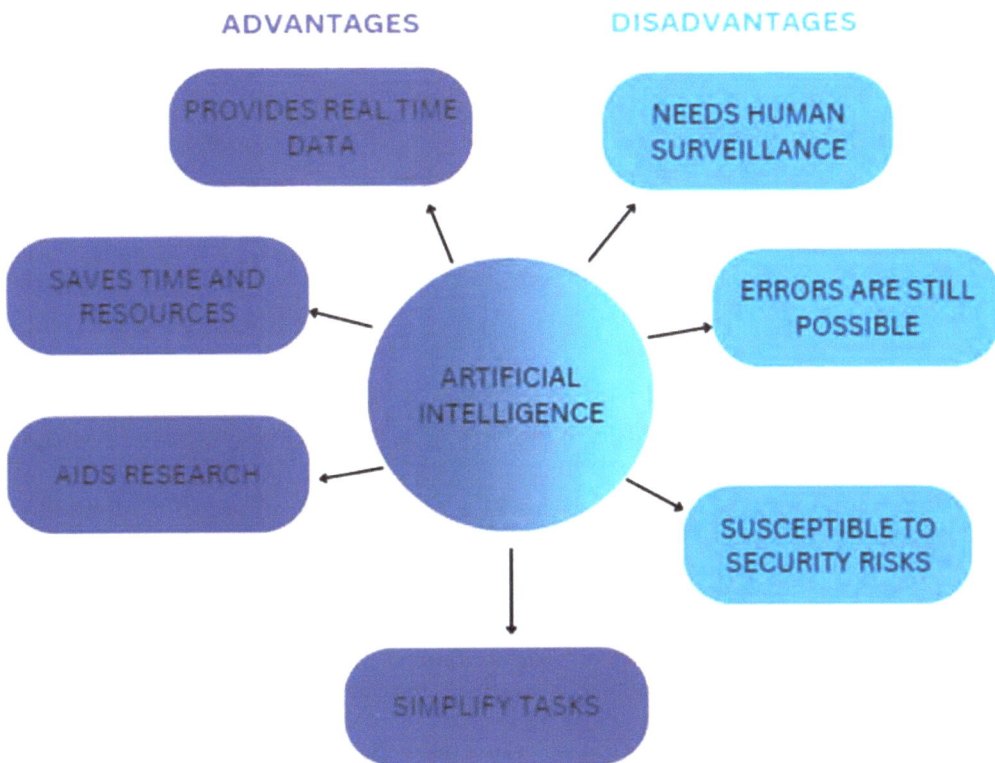

Fig. (4). Advantages and disadvantages of AI in medicine.

AI Applications in Healthcare

Stroke Research

- AI significantly contributes to stroke research, addressing a global burden of over 500 million cases.
- Increasing use of AI in prevention and treatment research, with stroke-related costs reaching $689 billion globally [38].

Health Management

- AI aids comprehensive health management by providing real-time updates on medical knowledge.
- Supports administrators, nurses, and medical staff, especially crucial during events like the Covid-19 pandemic [39].

Predictive Medicine

- AI assists in achieving diagnostic, therapeutic, and predictive outcomes by identifying meaningful patterns in raw data.
- Enables disease prevention, risk factor identification, and customization of treatment programs for improved patient outcomes [40].

Clinical Decision Making

- AI has the potential to enhance medical assessments, speed up procedures, and positively impact healthcare service costs.
- Algorithms provide valuable support for medical experts in making informed decisions.

Patient Diagnosis and Data

- AI manages vast amounts of patient data, optimizing processes like screening, diagnosis, and treatment allocation.
- Positive impact on surgery, physical therapy, and the development of rehabilitation robots [41].

PREVENTIVE HEALTH CARE

Preventive health care operates through layers consisting of a vast array of devices, technologies, and sensors interlinked through connected or wireless connections, including Wi-Fi connections.

On Body IoMT

The smartwatch comes with a wireless platform and an array of sensors designed for monitoring respiration rate, oxygen saturation, and other safety parameters. Similarly, a smartphone application system has been incorporated into the intelligent sensor-based outdoor jacket design, enabling the tracking of fall detection, emergency calls, heart rate, body temperature, and breathing rate. To promote the adoption of healthier behaviors, users have the option to receive real-time feedback and notifications concerning their health status.

Remote Patient Monitoring

The design of a system for monitoring patients from a distance prioritizes the immediate accuracy of electrocardiogram (ECG) readings. It employs the MQTT protocol for the swift dispatch of ECG data to an online server. This setup enables doctors to monitor ECG readings in real-time and access historical ECG data *via* web access on computers or smartphones. The system captures ECG readings through an AD8232 ECG sensor, which is adept at monitoring and enhancing the ECG signals that emanate from the cardiac muscle's electrical actions. Central to this design is the Arduino ESP 32 equipped with WiFi, which acts as the main processing unit. This unit processes ECG data as it occurs and forwards it through its WiFi component to the MQTT broker. A Raspberry Pi 3 hosts the MQTT broker, specifically the Mosquito version, streamlining the delivery of ECG readings to the online server.

Data Analytics and Predictive Modelling

Machine learning (ML) is revolutionizing the medical field by enabling robots to undertake vital and time-consuming tasks, particularly in medical imaging. Nowadays, machine learning (ML) plays a crucial role in identifying diseases and potential treatments. Healthcare data is considered paramount in ML and DL applications within healthcare systems. The potential of machine learning (ML) to enhance the efficiency and simplicity of medical treatment and care processes is significant. Moreover, it enhances the precision of predictions generated by ML algorithms, thus facilitating quicker decision-making by healthcare professionals.

Population Health Management

Beyond individual health analysis, the framework aggregates and analyses anonymized data from a broader population. This population health management approach allows for the identification of trends, risk factors, and common health challenges within specific demographic groups. Analyzing this data provides valuable insights into public health trends and informs policy decisions for healthcare organizations and authorities.

SENSORS USED IN THE INTERNET OF MEDICAL THINGS FOR PREVENTIVE HEALTHCARE

Sensors for Pulse Rate

Heart rate sensors, also known as pulse rate sensors, are crafted to measure a person's heart rate. Among the prevalent methods is Photoplethysmography (PPG), which leverages light to monitor fluctuations in blood volume. This technique typically involves emitting either red or green light onto the skin or fingertips, a portion of which is absorbed by blood vessels while the remainder is reflected to the sensor. Variations in blood circulation during heartbeats result in changes in light absorption and reflection, which are then captured by the sensor. These light alterations are subsequently transformed into electrical signals for processing. The processed data may be transmitted to a connected device for further analysis or pulse rate monitoring, or it can be displayed directly on a screen for immediate observation.

Pulse Oximeter

A pulse oximeter, widely utilized in medical environments, assesses an individual's blood oxygen saturation level (SpO_2). This compact, easy-to-carry device can be clipped onto body parts like the finger, earlobe, or toe. It operates by projecting two distinct wavelengths of light, red and infrared, through the skin where it's attached. The device then estimates the oxygen saturation by evaluating the oxygen-rich hemoglobin present in the blood, displaying the result as a percentage. Essential for tracking the health of patients with cardiovascular or lung diseases, pulse oximeters are also valuable for gauging the oxygen levels of people engaging in physical activities or those at elevated altitudes.

PPG Sensors

Photoplethysmography (PPG) sensors, integral to wearable devices like smartwatches and fitness trackers, are essential for measuring heart rate and blood oxygen levels non-invasively. By emitting light onto the skin and analyzing the reflected or transmitted light, PPG sensors can detect blood volume changes in the peripheral circulation, creating a waveform that reflects cardiovascular health. This technology enables continuous health monitoring, offering insights into an individual's wellness and physical condition without the need for medical intervention. As a cornerstone of health-related wearable technology, PPG sensors are pivotal for both personal wellness tracking and medical applications, facilitating a proactive approach to health management.

Blood Pressure Sensors

Blood pressure sensors rely on the force exerted by the circulation of blood within the arteries to gauge and track blood pressure levels. One common element present in contemporary blood pressure sensors is an oscillometer, designed to identify variations in pressure due to the rhythmic flow of blood. To observe pressure shifts, the sensor first applies pressure by inflating a cuff to momentarily impede blood flow, then slowly release it. Utilizing these data points, the sensor can ascertain both the systolic and diastolic measurements of blood pressure.

EMG Sensors

Electromyography (EMG) sensors are employed to capture the electrical signals that arise from muscle activity. Widely utilized across medical and scientific domains, these tools serve various purposes including the assessment of muscle functionality, the diagnosis of neuromuscular ailments, and the facilitation of prosthetic limb control Electromyography (EMG) sensors utilize electrodes positioned on the skin's surface, directly above the muscles being monitored. These electrodes detect the electrical activity generated during muscle contractions. When muscles are activated, signals are sent from the brain through nerves to stimulate specific muscle fibers, causing them to contract. The EMG sensors then capture these electrical signals and translate them into measurable data.

APPLICATION OF PREVENTIVE HEALTHCARE IN IOMT

The Haemodialysis Sensor Patch Identifies Blood Leaks

The system included a sensor with multiple rings to detect leaks as liquid volume rose, along with a Bluetooth low-energy module and a mapping circuit. During manufacturing, insulating layers kept a signal line, while only the sensing point was exposed. Blood leak detection involved absorption, response, and saturation stages: blood absorbed into the gauze diffused into the patch, causing voltage to rise, until saturation halted blood spread, maintaining a constant voltage level.

Advanced Medication for Hypertension

Monitoring doses will offer more accurate insights into treatment effectiveness tailored to everyone. Additionally, smart medications equipped with IoMT sensors embedded in each tablet can assess antibiotic levels in bodily fluids.

Digital Biomarkers

These devices facilitate the continuous monitoring of biomarkers found in bodily fluids like blood, urine, and sweat in real time. Consequently, distinguishing between infections caused by viruses and bacteria may become more straightforward, presenting numerous potential applications. Additionally, closely monitoring patients' responses to treatment becomes feasible, allowing assessment of treatment efficacy and the potential emergence of resistance.

Disease Surveillance and Tracking

The primary challenges in global disease management lie in testing and tracking, especially for the purpose of slowing transmission. Many IoMT-driven devices have been deployed to identify and monitor individuals afflicted with diseases, while also pinpointing patient locations to assess the risk of disease spread. Point-of-care testing (POCT) devices that utilize IoMT technology have the potential to improve the management of infectious diseases like malaria, dengue fever, influenza A (H1N1), human papillomavirus, Ebola virus disease, Zika virus, and coronavirus (COVID-19) with greater efficiency.

Stress and Anxiety Monitoring

The device is an affordable anxiety disorder monitor based on the Internet of Things (IoT), which utilizes physiological data within a semi-immersive environment to extract emotional characteristics. A Raspberry Pi 3 is employed to preprocess user data related to heart rate and physical activity before transferring it to an IoT cloud through an IoT node. Validation results of this system indicate a 90% accuracy rate in identifying anxiety disorders.

Seizure Detection

A rapid electrical disturbance in the brain is known as a seizure. Therefore, it is vital to promptly identify or recognize it to ensure patients receive appropriate treatment and care. By employing an IoMT-based approach, electroencephalography (EEG) data from a patient can be utilized to detect the onset of a seizure. The system continually processes and examines neurological data obtained from an EEG sensor to identify hyper-synchronous brain pulses. This extracted data is subsequently used to ascertain the frequency of seizures

Automated Insulin Injection

The Raspberry Pi 3, known for its affordability and versatility, serves as the foundation. A live video feed from a camera is employed to gather data on four essential bodily signs, which are then displayed on a website for continuous monitoring. Live monitoring is accessed through two distinct login credentials: one for the doctor and another for the patient's family. Any irregularity or instability detected by the sensors triggers a notification to the doctor *via* the GSM module. Upon determining the need for insulin injections, doctors can administer them remotely by activating a button that dispenses a controlled amount of insulin into the body.

TELEMEDICINE AND VIRTUAL HEALTHCARE SERVICE

Telemedicine refers to the delivery of medical care *via* interactive audiovisual and digital connections, encompassing services such as diagnosis, counselling, treatment, and the sharing of medical records and health education. With escalating healthcare costs and a growing need for improved care, many hospitals are exploring the benefits of telemedicine. They seek to optimize healthcare resources, improve communication between distant doctors and patients, and foster better connectedness, thereby reducing hospital readmissions and promoting patients' adherence to treatment plans.

IoT-based Telemedicine Demonstrations

- Remote Consultation: A physician guides an unwell patient who may require professional medical attention from a distance.
- Telemedical Assistance: A doctor aids another physician in conducting a therapeutic demonstration remotely.
- Distant Expertise: Physicians collaborate to evaluate and address issues from afar.
- Emergency Medical Coordination

IoT based Telemedicine Mobility

To ensure continuous medical care regardless of location, applications and networks for IoT-based telemedicine need to support patient mobility. This mobility aspect allows patients to access medical services both indoors and outdoors, which is particularly crucial for the elderly and individuals with serious conditions such as cardiovascular disease, diabetes, and endocrine and metabolic disorders, among others.

IoT-based Telemedicine for New Diseases

IoT is crucial for future telemedicine progress. While new telemedicine applications are emerging, they mainly target specific disease categories. It's vital to explore advanced data analysis methods like deep learning to extract accurate insights from complex datasets. Research on reporting erroneous symptoms is also essential to address emerging disorders. Utilizing sensors, devices, and innovative techniques facilitates the introduction of new telemedicine applications.

IoT-based Telemedicine Healthcare Services

Telemedicine clinical services encompass both synchronous and asynchronous healthcare services, with remote patient monitoring constituting one such service that consistently monitors patients' health status. Applications leveraging the Internet of Things for telemedicine necessitate a real-time operating system with additional stringent specifications.

IoT-Enabled Communication Tools

IoT-enabled communication solutions can improve in-the-moment interactions between patients and healthcare providers. The quick and easy implementation of healthcare consultations using messaging, video conferencing, and other communication tools.

Network Architectures

To facilitate data communication among the CMN, patients, and family doctors, a relatively high bandwidth is essential. Not all patients have access to broadband Internet; some may rely on low bandwidth options like ISDN or Modem connections. Given the mobile nature of the operation, wireless network technologies are necessary. Considering the extensive reach of physicians, the network must cover large distances while maintaining a consistent bandwidth of at least 3.0 Mbps.

WiMAX

The WiMAX point-to-multi-point configuration facilitates connections with multiple subscriber stations. For instance, in a healthcare setting, a WiMAX base station is established at a doctor's office. When nurses travel to attend to patients, they utilize portable WiMAX subscriber stations to establish connections with the base station. Additionally, within the coverage area, a WLAN-802.11g access point connects to the subscriber station. This setup enables communication between the nurse's Tablet PC, equipped with WLAN functionality, and the doctor's office PC.

PDA

The primary element of the positioning system is the PDA, which collects essential data for assessing WiMAX signal quality, determining the compass direction, and identifying geographic coordinates. Through a connection to the

WLAN access point, which receives WiMAX signal quality data, the PDA establishes a link with the subscriber station. Furthermore, the PDA connects to the GPS module *via* Bluetooth and utilizes a serial interface to interface with the compass system.

Applications of Telemedicine

Improving Doctor-Patient Interactions

IoT devices contribute to improving the quality of doctor-patient interactions in telemedicine.

The potential for increased patient engagement, better communication, and enhanced understanding of treatment plans using connected devices.

Personalized Medicine

AI enables the customization of treatment plans by analysing individual patient data, offering tailored therapies, and predicting responses based on genetic, lifestyle, and environmental factors.

The increasing occurrence of chronic illnesses presents a considerable hurdle for our healthcare infrastructure. Telemedicine software arises as an ideal remedy, streamlining and lowering the expenses associated with patients managing their health independently.

In a telehealth initiative tailored for individuals with congestive heart failure, following hospital discharge, there was a significant 73 percent reduction in 30-day hospital readmissions and a 50 percent decrease in readmissions within six months, highlighting the effectiveness of remote care.

Remote Chronic Disease Management

Managing chronic diseases remotely presents a significant challenge for our healthcare system due to their increasing prevalence. Telemedicine software emerges as a promising solution, offering patients a more convenient and cost-effective means to monitor and manage their health.

Remote Post-Hospitalization Care

Providing post-hospitalization care remotely has proven successful in various contexts. For instance, one telehealth program targeting congestive heart failure patients achieved a remarkable 73 percent reduction in 30-day hospital readmissions and a 50 percent decrease in six-month readmissions.

School-Based Telehealth

If a child falls ill at school, parents might opt to retrieve them and seek care at an urgent care facility or consult the school nurse. Some innovative school districts have collaborated with physicians to conduct house calls directly from the school premises. Besides offering parents advice or reassurance, the healthcare provider can assess the urgency of the situation.

Follow-Up Visits

Employing health software for regular follow-up appointments not only increases efficiency for both patients and clinicians but also opens the potential for improved follow-up, reduces the occurrence of missed appointments, and positively influences patient outcomes.

Future Applications of Telemedicine

Telemedicine facilitates easier access to healthcare services for patients facing challenges such as geographical distance (especially in rural regions), transportation limitations, or the unavailability of caregivers. It also ensures that patients with compromised immune systems no longer risk exposure to infectious diseases. Moreover, patients now can consult a variety of specialists nationwide, receiving prompter care compared to the months-long wait times for local specialists. In cases where patients unintentionally miss appointments due to such obstacles or mere forgetfulness, providers can utilize telemedicine to sustain treatment continuity, obviating the need for rescheduling and thereby minimizing time losses while enhancing clinic efficiency.

CHALLENGES FOR THE INTERNET OF MEDICAL THINGS (IOMT)

Simplified Connectivity

- An effective IoMT platform should provide straightforward and efficient access to IoT devices and data.
- Empowers developers to quickly create analytical applications, visualization dashboards, and IoMT applications.
- Facilitates easy device connection and management through scalable cloud-based services.

Efficient Device Management

Smart device management enhances asset availability, increases throughput, minimizes unplanned outages, and reduces maintenance costs. Various challenges in implementing the Internet of Medical Things are graphically represented in fig. (**5**).

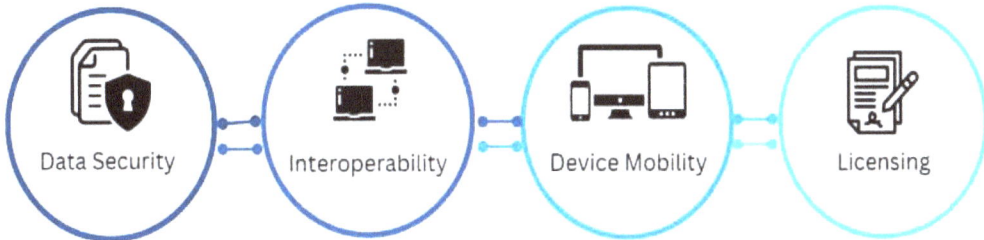

Fig. (5). IoMT challenges.

Data Ingestion

• IoMT platforms intelligently transform and store IoT data.
• Application Programming Interfaces (APIs) streamline the process of pulling necessary data from various sources and platforms. Rich analytics extract essential values from the data.

Insightful Analytics

• IoMT platforms should incorporate advanced analytics methods from the broader IoT domain.
• Real-time analytics enable responses to be evolving situations, learning from decisions made as circumstances change.
• An intuitive dashboard simplifies understanding of the analytics outcomes.

Key Capabilities for IoMT Platforms

These capabilities are crucial for addressing challenges associated with IoMT, ensuring seamless connectivity, effective device management, efficient data handling, and insightful analytics in the healthcare domain.

CONCLUSION

The Internet of Medical Things (IoMT) is revolutionizing healthcare by integrating advanced technologies to enhance patient care and improve health outcomes. This chapter has explored the diverse applications of IoMT, from wearable devices and surgical robots to telemedicine and preventive healthcare solutions. IoMT has ushered in an era of connected health devices, enabling real-time monitoring and personalized care. Wearable technologies and AI integration provide continuous health tracking and predictive analytics. Surgical robotics and cyber-physical health systems offer enhanced precision and efficiency in medical procedures. Telemedicine has emerged as a critical component, bridging geographical gaps and improving access to specialized care. However, challenges such as data security, interoperability, and ethical considerations must be addressed. As IoMT continues to evolve, it promises to create a more proactive, efficient, and personalized healthcare ecosystem. The integration of IoMT with emerging technologies will undoubtedly play a crucial role in shaping the future of medicine and public health.

REFERENCES

[1]   A.A. El-Saleh, "Abdul Manan Sheikh, Mahmoud, and Mohamed Shaik Honnurvali, "The Internet of Medical Things (IoMT): opportunities and challenges,"", *Wirel. Netw.,* no. May, 2024.
       [http://dx.doi.org/10.1007/s11276-024-03764-8]

[2]   C.H. Sumerli A, N.D. Erlinawati, D. Anurogo, D.M. Hasyim, and A. Ardi, "Application of the internet of things (IoT) in health: The future of personal care", *J. World Futur. Med. Health Nurs.,* vol. 2, no. 1, pp. 79-92, 2024.
       [http://dx.doi.org/10.70177/health.v2i1.705] [PMID: 38291421]

[3]   B. Bhushan, A. Kumar, A.K. Agarwal, A. Kumar, P. Bhattacharya, and A. Kumar, "Towards a secure and sustainable internet of medical things (IoMT): Requirements, design challenges, security techniques, and future trends", *Sustainability (Basel),* vol. 15, no. 7, p. 6177, 2023.
       [http://dx.doi.org/10.3390/su15076177]

[4]   A. Kadu, and M. Singh, "Comparative analysis of e-health care telemedicine system based on internet of medical things and artificial intelligence", *2nd International Conference on Smart Electronics and Communication (ICOSEC),* pp. 1768-1775, 2021.
       [http://dx.doi.org/10.1109/ICOSEC51865.2021.9591941]

[5]   S.A. Ajagbe, J.B. Awotunde, A.O. Adesina, P. Achimugu, and T.A. Kumar, "Internet of medical things (IoMT): Applications, challenges, and prospects in a data-driven technology", *Intelligent Healthcare,* pp. 299-319, 2022.
       [http://dx.doi.org/10.1007/978-981-16-8150-9_14]

[6]   A. Aliverti, "Wearable technology: role in respiratory health and disease", *Breathe (Sheff.),* vol. 13, no. 2, pp. e27-e36, 2017.
       [http://dx.doi.org/10.1183/20734735.008417] [PMID: 28966692]

[7]   X. Wang, "The architecture design of the wearable health monitoring system based on internet of things technology", *Int. J. Grid Util. Comput.,* vol. 6, no. 3/4, p. 207, 2015.
       [http://dx.doi.org/10.1504/IJGUC.2015.070681]

[8]   C. Dinh-Le, R. Chuang, S. Chokshi, and D. Mann, "Wearable health technology and electronic health record integration: Scoping review and future directions", *JMIR Mhealth Uhealth,* vol. 7, no. 9, p. e12861, 2019.
       [http://dx.doi.org/10.2196/12861] [PMID: 31512582]

[9]   S. DiMaio, M. Hanuschik, and U. Kreaden, "The da vinci surgical system", *Surgical Robotics,* no. Nov, pp. 199-217, 2011.
       [http://dx.doi.org/10.1007/978-1-4419-1126-1_9]

[10]   S. Takahashi, E. Nakazawa, S. Ichinohe, A. Akabayashi, and A. Akabayashi, "Wearable technology for monitoring respiratory rate and SpO$_2$ of COVID-19 patients: A systematic review"., *Diagnostics (Basel),* vol. 12, no. 10, p. 2563, 2022.
       [http://dx.doi.org/10.3390/diagnostics12102563] [PMID: 36292252]

[11]   B.S. Peters, P.R. Armijo, C. Krause, S.A. Choudhury, and D. Oleynikov, "Review of emerging surgical robotic technology", *Surg. Endosc.,* vol. 32, no. 4, pp. 1636-1655, 2018.
       [http://dx.doi.org/10.1007/s00464-018-6079-2] [PMID: 29442240]

[12]   M. Kujawska, S.K. Bhardwaj, Y.K. Mishra, and A. Kaushik, "Using graphene-based biosensors to detect dopamine for efficient parkinson's disease diagnostics", *Biosensors (Basel), vol.* 11, no. 11, p. 433, 2021.
[http://dx.doi.org/10.3390/bios11110433] [PMID: 34821649]

[13]   P. Manickam, S.A. Mariappan, S.M. Murugesan, S. Hansda, A. Kaushik, R. Shinde, and S.P. Thipperudraswamy, "Artificial intelligence (AI) and internet of medical things (IoMT) assisted biomedical systems for intelligent healthcare", *Biosensors (Basel), vol.* 12, no. 8, p. 562, 2022.
[http://dx.doi.org/10.3390/bios12080562] [PMID: 35892459]

[14]   S.A. Haque, S.M. Aziz, and M. Rahman, "Review of cyber-physical system in healthcare", *Int. J. Distrib. Sens. Netw.,* vol. 10, no. 4, p. 217415, 2014.
[http://dx.doi.org/10.1155/2014/217415]

[15]   N.M. Thomasian, and E.Y. Adashi, "Cybersecurity in the Internet of Medical Things", *Health Policy Technol.,* vol. 10, no. 3, p. 100549, 2021.
[http://dx.doi.org/10.1016/j.hlpt.2021.100549]

[16]   A. Chandrashekhar, R. Raffik, R. Sridevi, M. Sindhu, K. Rajkumar, and T. Jaiswal, "3D-Printed human organ designs with tissue physical characteristics and embedded sensors," pp. 135–152, Feb. 2024,
[http://dx.doi.org/10.1002/9781394197705.ch9]

[17]   M. Osama, A.A. Ateya, M.S. Sayed, M. Hammad, P. Pławiak, A.A. Abd El-Latif, and R.A. Elsayed, "Internet of medical things and healthcare 4.0: Trends, requirements, challenges, and research directions", *Sensors (Basel),* vol. 23, no. 17, p. 7435, 2023.
[http://dx.doi.org/10.3390/s23177435] [PMID: 37687891]

[18]   G. Aceto, V. Persico, and A. Pescapé, "Industry 4.0 and health: Internet of things, big data, and cloud computing for healthcare 4.0", *J. Ind. Inf. Integr.,* vol. 18, p. 100129, 2020.
[http://dx.doi.org/10.1016/j.jii.2020.100129]

[19]   H.T. Yew, M.F. Ng, S.Z. Ping, S.K. Chung, A. Chekima, and J.A. Dargham, "IoT based real-time remote patient monitoring system", *16th IEEE International Colloquium on Signal Processing & Its Applications (CSPA),* pp. 176-179, 2020.
[http://dx.doi.org/10.1109/CSPA48992.2020.9068699]

[20]   L. Linkous, N. Zohrabi, and S. Abdelwahed, "Health monitoring in smart homes utilizing internet of things", *IEEE/ACM International Conference on Connected Health: Applications, Systems and Engineering Technologies (CHASE),* pp. 29-34, 2019.
[http://dx.doi.org/10.1109/CHASE48038.2019.00020]

[21]   S.M.N. Islam, "The role of the internet of things in healthcare transformation", *Int. Trans. Artif. Intell. (ITALIC),* vol. 2, no. 1, pp. 1-6, 2023.
[http://dx.doi.org/10.33050/italic.v2i1.380]

[22]   P.P. Ray, D. Dash, and N. Kumar, "Sensors for internet of medical things: State-of-the-art, security and privacy issues, challenges and future directions", *Comput. Commun.,* vol. 160, pp. 111-131, 2020.
[http://dx.doi.org/10.1016/j.comcom.2020.05.029]

[23]   P. Wal, A. Wal, N. Verma, R. Karunakakaran, and A. Kapoor, "Internet of medical things – the future of healthcare", *Open Public Health J.,* vol. 15, no. 1, p. e187494452212150, 2022.
[http://dx.doi.org/10.2174/18749445-v15-e221215-2022-142]

[24]   S. Chen, L. Jiang, L. Wu, Y. Wang, and A. Usmani, "Damage investigation of cementitious fire resistive coatings under complex loading", *Construction and Building Materials,* vol. 204, pp. 659-674, 2019.

[25]   F. Subhan, A. Mirza, M.B.M. Su'ud, M.M. Alam, S. Nisar, U. Habib, and M.Z. Iqbal, "AI-enabled wearable medical internet of things in healthcare system: A survey", *Appl. Sci. (Basel),* vol. 13, no. 3, p. 1394, 2023.

[http://dx.doi.org/10.3390/app13031394]

[26]   Aminu Muhammad Auwal, "IoT integration in telemedicine: Investigating the role of internet of things devices in facilitating remote patient monitoring and data transmission", *Research Square*, 2023.
[http://dx.doi.org/10.21203/rs.3.rs-3419693/v1]

[27]   M. Khoshnam, M. Azizian, and R.V. Patel, "Modeling of a steerable catheter based on beam theory", *2012 IEEE International Conference on Robotics and Automation*, pp. 4681-4686.
[http://dx.doi.org/10.1109/ICRA.2012.6224784]

[28]   I.K. Abdelghany, R. AlMatar, A. Al-Haqan, I. Abdullah, and S. Waheedi, "Exploring healthcare providers' perspectives on virtual care delivery: insights into telemedicine services", *BMC Health Serv. Res.*, vol. 24, no. 1, p. 1, 2024.
[http://dx.doi.org/10.1186/s12913-023-10244-w] [PMID: 38169381]

[29]   Z. Islam, M.R.I. Bhuiyan, T.A. Poli, R. Hossain, and L. Mani, "Gravitating towards internet of things: Prospective applications, challenges, and solutions of using IoT", *Int. J. Relig.*, vol. 5, no. 2, pp. 436-451, 2024.
[http://dx.doi.org/10.61707/awg31130]

[30]   V. Chaudhary, A. Kaushik, H. Furukawa, and A. Khosla, "Review—Towards 5th Generation AI and IoT driven sustainable intelligent sensors based on 2D MXenes and borophene"., *ECS Sensors Plus*, vol. 1, no. 1, p. 013601, 2022.

[31]   M. Krohn, H. Kopp, and D. Tavangarian, "A wireless architecture for telemedicine", *4th Workshop on Positioning, Navigation and Communication*, pp. 109-111, 2007.
[http://dx.doi.org/10.1109/WPNC.2007.353620]

[32]   F. Troyes, "Magnetic nanoparticle-based biosensors for the sensitive and selective detection of urine glucose", *Recent Advancements in Biomedical Engineering*, p. 21, 2022.

[33]   R. Cao, Z. Tang, C. Liu, and B. Veeravalli, "A Scalable multicloud storage architecture for cloud-supported medical internet of things", *IEEE Internet Things J.*, vol. 7, no. 3, pp. 1641-1654, 2020.
[http://dx.doi.org/10.1109/JIOT.2019.2946296]

[34]   D. Ganesh, G. Seshadri, S. Sokkanarayanan, P. Bose, S. Rajan, and M. Sathiyanarayanan, "Automatic health machine for covid-19 and other emergencies", *2021 International Conference on COMmunication Systems & NETworkS (COMSNETS)*, IEEE., pp. 685-689, 2021.

[35]   V. Marda, "Artificial intelligence policy in India: A framework for engaging the limits of data-driven decision-making", *Philos. Trans.- Royal Soc., Math. Phys. Eng. Sci.*, vol. 376, no. 2133, p. 20180087, 2018.
[http://dx.doi.org/10.1098/rsta.2018.0087] [PMID: 30323001]

[36]   M.S. Adhikari, P.S. Khan, G. Senapathi, B. Chatterjee, D.V. Reddy, and P.K. Malik, "Design of an IoT based smart medicine box", *2023 IEEE Devices for Integrated Circuit (DevIC)*, IEEE., pp. 346-349, 2023.

[37]   S. Dogra, L. Singh, and A. Gupta, "Low–cost portable smart ventilator", *Recent Advances in Manufacturing, Automation, Design and Energy Technologies: Proceedings from ICoFT*, Springer Singapore., pp. 599-606, 2022.

[38]   B. Wahl, A. Cossy-Gantner, S. Germann, and N.R. Schwalbe, "Artificial intelligence (AI) and global health: How can AI contribute to health in resource-poor settings?", *BMJ Glob. Health*, vol. 3, no. 4, p. e000798, 2018.
[http://dx.doi.org/10.1136/bmjgh-2018-000798] [PMID: 30233828]

[39]   A.K. Saenger, and R.H. Christenson, "Stroke biomarkers: Progress and challenges for diagnosis, prognosis, differentiation, and treatment", *Clin. Chem.*, vol. 56, no. 1, pp. 21-33, 2010.
[http://dx.doi.org/10.1373/clinchem.2009.133801] [PMID: 19926776]

[40]   E. Heeley, C.S. Anderson, Y. Huang, S. Jan, Y. Li, M. Liu, J. Sun, E. Xu, Y. Wu, Q. Yang, J. Zhang,

S. Zhang, and J. Wang, "Role of health insurance in averting economic hardship in families after acute stroke in China", *Stroke,* vol. 40, no. 6, pp. 2149-2156, 2009.
[http://dx.doi.org/10.1161/STROKEAHA.108.540054] [PMID: 19359646]

[41]   B.X. Tran, G.T. Vu, G.H. Ha, Q.H. Vuong, M.T. Ho, T.T. Vuong, V.P. La, M.T. Ho, K.C.P. Nghiem, H.L.T. Nguyen, C.A. Latkin, W.W.S. Tam, N.M. Cheung, H.K.T. Nguyen, C.S.H. Ho, and R.C.M. Ho, "Global evolution of research in artificial intelligence in health and medicine: A bibliometric study", *J. Clin. Med.,* vol. 8, no. 3, p. 360, 2019.
[http://dx.doi.org/10.3390/jcm8030360] [PMID: 30875745]

CHAPTER 5

Decoding Diabetic Retinopathy Images: A Comparative Analysis of Deep Learning Models for Effective Severity Grading

Soumya Ranjan Mahanta[1,*]

[1] *Department of Computer Science, Utkal University, Bhubaneswar, Odisha, India*

Abstract: Diabetic retinopathy (DR) is a leading cause of vision loss, making early detection and accurate severity assessment crucial for timely intervention. This study evaluated five pre-trained deep-learning models for classifying DR severity levels from fundus images. Encompassing diverse architectures like ResNet50 and EfficientNetB5, the models were extensively trained and validated for generalizability. Performance was assessed using ordinal regression metrics (MAE, MSE, QWK), ROC-AUC, specificity, sensitivity, and visualizations like confusion matrices and model architectures.

DenseNet121 and EfficientNetB5 emerged as top performers, with DenseNet121 excelling in precise severity predictions due to its outstanding MAE, MSE, and QWK scores. Confusion matrices provided insights into misclassifications, while ROC-AUC analysis confirmed the models' strong discriminative ability. DenseNet121 and EfficientNetB5 hold significant promise for aiding early DR detection and accurate severity assessment, potentially paving the way for timely intervention and personalized patient care. Future research should focus on fine-tuning strategies and external validation to enhance model generalizability and real-world clinical utility.

Keywords: Diabetic Retinopathy (DR), Deep Learning (DL), DenseNet121, EfficientNetB5, Early detection, Personalized care, Severity classification.

INTRODUCTION

Diabetic retinopathy (DR), a microvascular complication arising from diabetes mellitus, remains a significant global public health concern, consistently ranked as a leading cause of vision loss. Early and accurate diagnosis plays a pivotal role in enabling timely intervention and preventing irreversible vision impairment. Recent advancements in deep learning (DL) techniques, particularly those

* **Corresponding author Soumya Ranjan Mahanta:** Department of Computer Science, Utkal University, Bhubaneswar, Odisha, India; E-mail: dipusoumyaranjan019@gmail.com

Parikshit N. Mahalle, Gitanjali R. Shinde, Namrata N. Wasatkar & Prashant R. Anerao (Eds.)

leveraging deep neural networks (DNNs), have reshaped the landscape of medical image analysis, offering immense promise for automated DR detection.

This chapter embarks on a comprehensive review of the relevant literature, specifically focusing on methodologies that employ DNNs, particularly convolutional neural networks (CNNs) and image-based DL approaches, for automated DR diagnosis. The exploration begins by examining the groundbreaking work [1] introducing Densely Connected Convolutional Networks (DenseNets). This novel architecture demonstrably enhanced information flow within neural networks, improving learning capabilities. Their work catalyzed subsequent research endeavors [2 - 4] that explored the potential of CNNs in achieving heightened accuracy and sensitivity in DR diagnosis. Recognizing the need for further optimization, researchers have actively pursued advancements in existing DL architectures specifically tailored for DR detection. A study proposed a feature-based optimized deep residual network architecture, improving model performance [5]. The introduction of the EfficientNet model [6] further revolutionized CNN scaling, achieving superior performance while maintaining computational efficiency. These advancements significantly contribute to the potential for practical DL application in real-world clinical settings.

This chapter extends beyond a mere examination of CNN architectures. Valuable insights from a broader spectrum of research [7 - 10] are incorporated. These studies delve into critical aspects that are essential for the field's continued progress, encompassing:

• Seamless integration of DL into robust and automated DR detection systems, facilitating broader clinical adoption.
• Development of reliable algorithms ensuring accurate and generalizable diagnosis across diverse patient populations.
• Comprehensive surveys that map the DL landscape for DR classification, foster a more holistic understanding of the field and its potential applications.

However, a discernible research gap emerges in the exploration of deep image mining techniques for efficient and comprehensive DR screening. The pioneering study introduces this concept, hinting at its potential to enhance sensitivity and specificity in DR screening [11]. Further investigations and developments are warranted to harness the potential of deep image mining approaches, potentially pushing the boundaries of what is achievable in automated diabetic retinopathy diagnosis.

This chapter delves into the world of deep learning (DL) for diabetic retinopathy (DR) detection. By drawing on insights from various studies, it dissects current methods, pinpoints emerging trends, and proposes exciting directions for future exploration. Ultimately, this chapter aims to significantly contribute to the development of accurate, efficient, and accessible tools for early DR detection. The goal is to significantly contribute to the development of accurate, efficient, and accessible tools for early DR detection. This, in turn, has the potential to improve patient outcomes and substantially reduce the global burden of vision loss associated with DR. While the current study explored ResNet, DenseNet, and EfficientNet architectures, further investigation is necessary to identify the optimal choice for real-world clinical settings. External validation with diverse datasets, fine-tuning hyperparameters specifically for DR tasks, and exploring multi-class classification alongside ordinal regression are all crucial for achieving robust and generalizable DR diagnosis. These future directions will solidify the most appropriate deep-learning architecture for accurate DR detection.

RELATED WORK

Diabetic retinopathy (DR) remains a significant global public health concern. Early detection through effective screening programs is critical for preventing vision loss. Deep learning has emerged as a powerful tool for automated DR analysis, offering promising avenues for both detection and severity grading. This section reviews relevant research on deep learning for DR image analysis, focusing specifically on approaches for severity grading. Diabetic retinopathy (DR) detection using deep learning (DL) has burgeoned into a dynamic and rapidly evolving field, marked by impactful research endeavors. This section conducts a critical examination and categorization of the existing literature, providing valuable insights into the development and application of DL techniques for DR classification.

Atwany, Sahyoun, and Yaqub (2022) present a comprehensive survey that serves as a cornerstone reference for the field. Their meticulous work maps the landscape of available DL methodologies for DR classification, establishing a robust foundation for further research endeavors [12]. This foundational survey provides a systematic overview, guiding subsequent research in understanding the intricacies of DL techniques for diabetic retinopathy detection. Kermany, Goldbaum, Cai, Valentim, Liang, Baxter, *et al.* (2018) laid the groundwork for image-based DL in medical diagnoses, showcasing its remarkable potential for identifying treatable diseases, including diabetic retinopathy. While extending beyond DR, their work emphasizes the broader impact of DL across healthcare applications, paving the way for its integration into various medical domains [13]. This pioneering research underscores the transformative potential of image-based DL not only for diabetic retinopathy but also for a wide range of medical conditions, driving advancements in healthcare technology.

Wang, Zhu, Wang, and Zhang (2021) offer a broad perspective on the burgeoning field of DL in medical imaging, while Suganyadevi, Seethalakshmi, and Balasamy (2022) delve specifically into the nuances of DL within this domain, providing a more focused understanding of its intricacies [14, 15]. These comprehensive reviews shed light on the diverse applications of DL in medical imaging, offering valuable insights into its potential to advance diagnostic capabilities across various healthcare domains. Kaushik, Singh, Kaur, Alshazly, Zaguia, and Hamam (2021) propose a novel approach for DR diagnosis using stacked generalization, strategically combining predictions from multiple DL models. This work demonstrates the potential of ensemble learning to significantly enhance diagnostic accuracy, potentially leading to more robust and reliable detection of DR [16]. The utilization of ensemble learning techniques highlights innovative strategies for improving the performance of DL models in diabetic retinopathy diagnosis, offering promising avenues for enhancing clinical decision-making processes. Alyoubi, Shalash, and Abulkhair (2020) and Lam, Yi, Guo, and Lindsey (2018) contribute systematic reviews that not only evaluate DL techniques for DR detection but also explore the potential integration of DL into automated systems. This line of inquiry delves into facilitating efficient clinical adoption, paving the way for the practical implementation of DL-based DR detection tools in real-world healthcare settings [17, 18]. Through systematic evaluation and integration into clinical practice, these reviews bridge the gap between DL research and real-world application, fostering the adoption of DL-based solutions for diabetic retinopathy detection in clinical settings. Burlina, Paul, Mathew, Joshi, Pacheco, and Bressler (2020) explore low-shot deep learning for DR, addressing the challenge of limited training data, a common hurdle in real-world applications [19]. Tsiknakis, Theodoropoulos, Manikis, Ktistakis, Boutsora, Berto, *et al.* (2021) investigate the application of DL for rare ophthalmic diseases, demonstrating its potential to address bias in retinal diagnostics, promoting fairer and more generalizable diagnostic tools [20]. These innovative approaches not only tackle specific challenges in diabetic retinopathy detection but also contribute to the broader field of medical imaging, offering solutions that pave the way for more inclusive and effective diagnostic tools.

Mehboob, Akram, Alghamdi, and Abdul Salam (2022) propose a deep learning-based approach for grading diabetic retinopathy, highlighting the significance of large image datasets. Their work contributes to refining diagnostic precision through advanced DL models, potentially leading to more nuanced and accurate grading of DR severity [21]. Addressing the crucial aspect of grading, their research enhances the granularity of diagnostic outcomes, facilitating a more precise understanding of the severity of diabetic retinopathy. Gulshan, Peng, Coram, Stumpe, Wu, Narayanaswamy, *et al.* (2016) and Varanasi and Dasari (2022) focus on the development and validation of DL algorithms specifically tailored for DR detection. Their meticulous work underscores the importance of robust models for reliable application in clinical settings, ensuring the trustworthiness and efficacy of DL-based tools for DR diagnosis [22, 23]. The emphasis on development and validation aligns with the imperative of translating DL advancements into clinically viable tools, fostering confidence in their application within medical practice. Gondal, Köhler, Grzeszick, Fink, and Hirsch (2017) and Al-Mukhtar, Morad, Albadri, and Islam (2021) explore weakly supervised localization techniques for diabetic retinopathy lesions. These works address the challenge of identifying lesions without extensive manual annotations, a time-consuming and resource-intensive process. Their research paves the way for more efficient training of DL models for DR detection [24, 25]. By addressing the labor-intensive nature of lesion annotation, these studies streamline the training process, enhancing the feasibility of deploying DL models in real-world clinical scenarios. Naithani, Bharadwaj, and Kumar (2019) explore automated detection frameworks and weakly supervised heatmap generation, suggesting promising avenues for future research and development in diabetic retinopathy detection using deep learning [26]. The investigation into automated detection frameworks and innovative heatmap generation techniques signifies a forward-looking approach, indicating the potential for breakthroughs that could shape the future landscape of automated diabetic retinopathy diagnosis.

GAP ANALYSIS

While the analysis effectively highlights DenseNet-121 as a potential leading performer in DR severity detection, critical gaps remain unresolved. Firstly, there's a notable deficiency in discussing methods to enhance model interpretability, which impedes clinician's understanding of predictions. Additionally, the insufficient exploration of potential future research directions beyond ensemble methods, including novel model architectures or the incorporation of additional clinical data modalities, highlights a gap in the discussion. Moreover, the limited exploration of future research avenues beyond ensemble approaches, such as novel model architectures or additional clinical data modalities, further underscores an inadequacy in the discussion. Lastly, the insufficient discourse on practical model deployment challenges, like regulatory compliance and integration with existing healthcare infrastructure, signifies a need for more comprehensive consideration of real-world applicability.

DATASET

The analysis leveraged the publicly available APTOS 2019 Blindness Detection dataset hosted on Kaggle [27]. This dataset encompasses nearly 1,928 de-identified retinal fundus photographs acquired from rural patient populations in

India. Each image was meticulously graded by a clinician on a five-point scale to indicate the severity of DR. Fig. (**1**) showcases sample images from the dataset. Notably, the dataset reflects real-world conditions, incorporating inherent variations in image quality. These variations encompass artifacts, blurriness, and inconsistencies in exposure due to the utilization of multiple cameras across various clinics over an extended period. This real-world characteristic of the data bolsters the generalizability of any model developed using this dataset for DR detection and severity assessment. Ultimately, the aim is to develop models capable of early DR detection, potentially preventing vision loss before it irreversibly progresses.

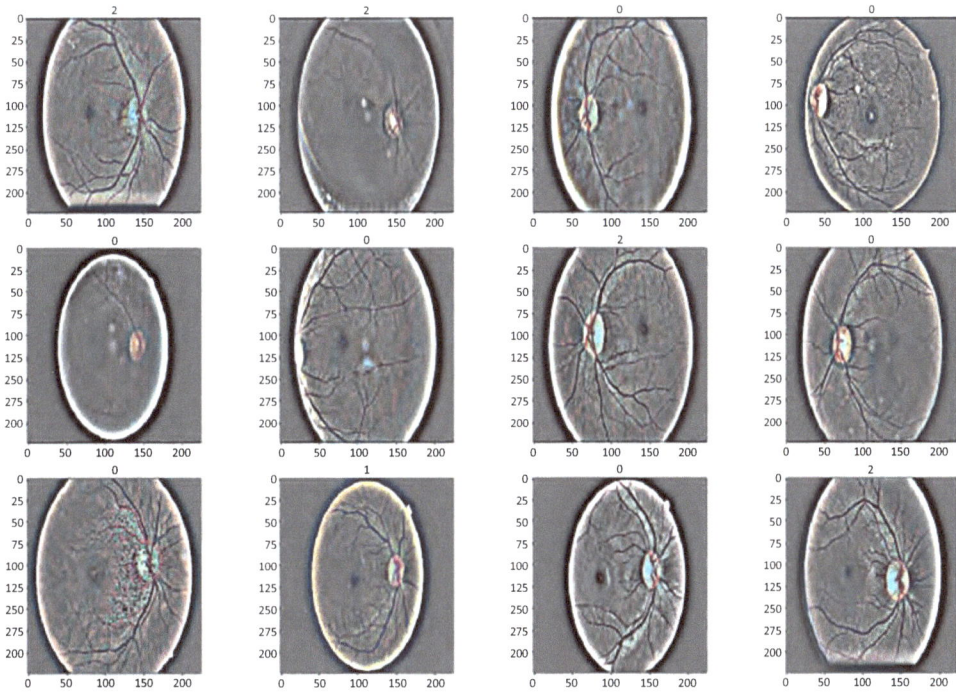

Fig. (1). Sample images from the dataset.

Descriptive Statistics

The dataset comprises retina images graded by clinicians for diabetic retinopathy severity on a scale ranging from 0 to 4, with categories as follows: (0: No DR, 1: Mild, 2: Moderate, 3: Severe, 4: Proliferative DR). Images exhibit common real-world noise elements such as artifacts, blurring, and exposure inconsistencies. Collected from various clinics over time and using different cameras, they introduce additional variability.

File attributes may vary between the public and private test sets in synchronous Kernels-only datasets, encompassing differences in IDs, sizes, and other parameters. Code should generate predictions adhering to the format specified in public sample_submission.csv without incorporating fixed aspects. During private testing, Kaggle substitutes the private test set and sample submission files, with the private test set estimated to comprise approximately 13,000 images, totaling 20GB.

Evaluation Metric

Within the field of diabetic retinopathy (DR) severity grading, the selection of an appropriate evaluation metric is crucial for the accurate assessment of model performance. Accuracy, a prevalent metric in classification tasks, can be deceptive when applied to imbalanced datasets, a common characteristic of DR data. In such scenarios, a model might achieve high accuracy by simply predicting the majority class, regardless of its ability to discern between distinct severity levels [28].

To address this limitation, the research community has adopted Quadratic Weighted Kappa (QWK), also known as Cohen's Kappa, as the standard evaluation metric for DR severity grading competitions. QWK transcends the simplistic measurement of agreement between predictions and ground truth labels. It incorporates a penalty for models that assign labels purely by chance, offering a more robust evaluation framework.

The QWK formula can be deconstructed as follows:

$$K \equiv \frac{po - pe}{1 - p} \tag{1}$$

- K (kappa) represents the overall agreement between the model's predictions and the ground truth labels.
- *po* (observed agreement) reflects the proportion of times the model and human experts agree on the severity level. It is analogous to accuracy but considers all classes, not just the majority class.
- *pe* (chance agreement) signifies the hypothetical probability that two observers would agree by pure chance, considering the class distribution of the data. This is calculated using the observed data itself.

By factoring in both observed agreement (*po*) and the likelihood of chance agreement (*pe*), QWK provides a more nuanced evaluation metric compared to accuracy. It penalizes models that simply predict the majority class and rewards models that can accurately differentiate between different severity levels. In essence, QWK offers a more sophisticated evaluation metric compared to accuracy. This makes it the preferred choice for assessing the effectiveness of DR severity grading models, particularly when dealing with imbalanced datasets, ensuring a more reliable and informative evaluation process.

PROPOSED METHODOLOGY

The chapter on diabetic retinopathy (DR) detection through deep learning initiates with the data collection and preprocessing phase. This involves sourcing high-quality retinal images from established datasets, including APTOS 2019, EyePACS 15, and Messidor-2, ensuring diversity to cover various DR severities. Subsequently, the collected data undergoes division into training, validation, and testing sets for foundational model learning, hyperparameter tuning, and objective assessment of model generalization, respectively. Preprocessing techniques encompass optional image cropping, standardization through resizing, optional data augmentation using Keras ImageDataGenerator, and multi-label encoding for comprehensive severity class prediction. Table 1 provides an overview of deep learning architectures and their performance considerations, aiding in informed decision-making during the model selection process. Transitioning to the deep learning model development phase, architecture selection involves choosing a pre-trained model such as ResNet, EfficientNet, or DenseNet-121. This model is tailored to DR detection needs by modifying final layers, incorporating global average pooling, dropout layers, and a dense output layer for severity score prediction. The mean squared error (MSE) loss function and Adam optimizer are employed for training, complemented by a custom Quadratic Weighted Kappa (QWK) score callback to monitor model performance and facilitate model selection based on the highest achieved QWK score.

Table 1. Deep learning architectures: overview and performance considerations.

Architecture	Working Principle	Dependencies Affecting Performance	Considerations for Optimal Performance Tuning
ResNet50	Addresses vanishing gradient by introducing skip connections	☐ Variant Selection: Consider the trade-off between depth and computational resources. ☐ Hyperparameter Tuning: Rigorous tuning of the learning rate, epochs, and optimizer is crucial for optimal convergence and generalization. ☐ Evaluation Strategy: Use a well-defined validation set to prevent overfitting.	☐ Carefully select the ResNet variant based on resource constraints. ☐ Perform hyperparameter tuning using grid search or random search. Evaluate the model on a dedicated validation set.

(Table 1) cont.....

Architecture	Working Principle	Dependencies Affecting Performance	Considerations for Optimal Performance Tuning
EfficientNet	Utilizes compound scaling for efficiency and high accuracy	☐ Variant Selection: Choose a variant (*e.g.*, EfficientNet-B5) balancing performance and efficiency. ☐ Dataset Size and Augmentation: Requires a larger dataset, consider augmentation for limited data.	☐ Choose the EfficientNet variant based on dataset size and resources. ☐ Leverage data augmentation for improved model generalization. ☐ Explore transfer learning for faster convergence and enhanced performance.
DenseNet	Employs dense connectivity for improved gradient flow	☐ Hyperparameter Tuning: Tune growth rate for optimal configuration. ☐ Data Augmentation: Use techniques for improved generalization, especially in imbalanced datasets.	☐ Carefully tune the growth rate to prevent overfitting. ☐ Utilize data augmentation for a diverse training set.

Fig. (**2**) illustrates the Methodology Flowchart for Diabetic Retinopathy Detection. The training strategy incorporates bucket-wise training for large datasets and implicit early stopping based on Quadratic Weighted Kappa (QWK) score improvement. The evaluation comprises performance visualization through line plots depicting metrics like loss, mean absolute error (MAE), and QWK against the number of epochs. Model testing on the unseen testing set follows, utilizing metrics such as accuracy, precision, recall, and F1-score to assess effectiveness in real-world DR detection scenarios.

An OptimizedRounder class is introduced for the optimized rounding of predicted continuous scores into discrete class labels representing DR severity levels. This class employs optimization techniques like Nelder-Mead to maximize the QWK score on the training data, ensuring robust mapping between predicted scores and clinically relevant DR classifications for enhanced interpretability and usability in clinical settings.

Fig. (**3**) provides an overall structured methodology for data collection, preprocessing, deep learning model development, evaluation, and optimized rounding. This framework offers a comprehensive approach to assess and refine the performance of deep learning models in diabetic retinopathy detection, contributing to advancements in clinical diagnosis and patient care.

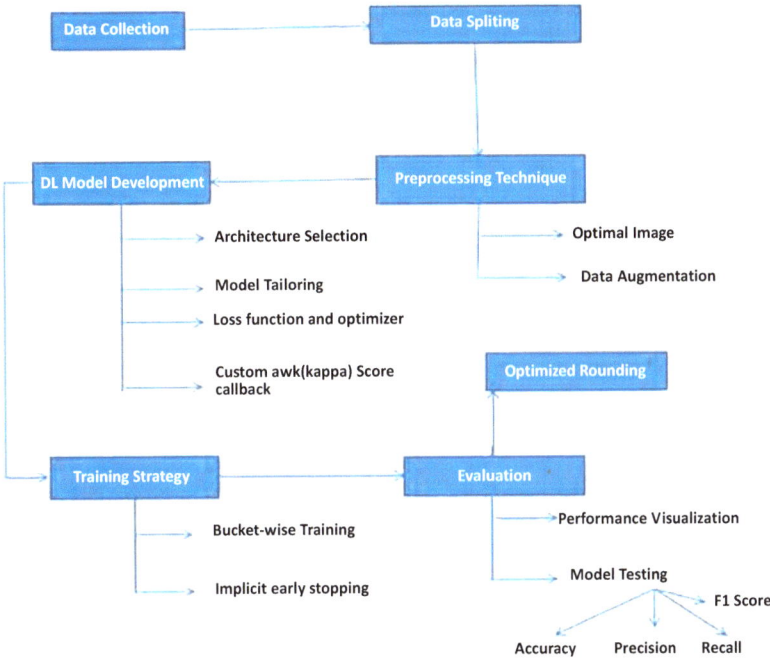

Fig. (2). Methodology flowchart for diabetic retinopathy detection.

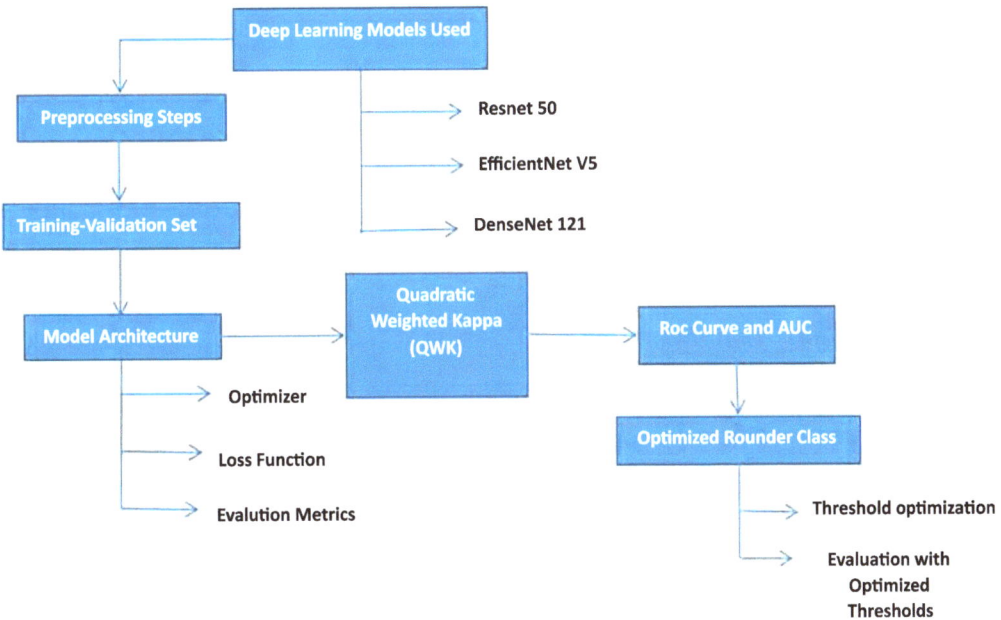

Fig. (3). Deep learning models diagnosis in diabetic retinopathy detection.

EXPERIMENTS AND EVALUATIONS

The three deep learning models EfficientNetB5, ResNet50, and DenseNet-121 were optimized and evaluated using metrics such as QWK score, accuracy, and model coefficients. The analysis identified the most effective model for the application.

Experimental Analysis of EfficientNetB5 for Diabetic Retinopathy Detection

This comprehensive experimental analysis explores the development, training, and evaluation of a deep learning model designed for the detection of diabetic retinopathy (DR) using fundus images. The meticulous approach encompasses several key stages, including data preparation, model architecture, training details, and evaluation metrics.

Pre-processing Steps

The initial phase involves critical pre-processing steps to optimize the fundus images for the model. This includes meticulous cropping to focus on the relevant region of interest, resizing for standardized dimensions compatible with pre-trained models, and normalization to scale pixel intensities. The chosen techniques and their justifications, such as employing a fixed-size square crop centered on the optic nerve, are thoroughly detailed.

Model Architecture

The model is based on the EfficientNetB5 variant, specifically (EfficientNetB5-Base), chosen for efficiency and performance. Custom layers, including fully connected layers and a regression output layer, are strategically incorporated to adapt the model for the DR severity regression task, with detailed descriptions of their functionalities.

Model Execution - Training Details

The model's execution involves comprehensive training details. The Adam optimizer is employed, and hyperparameters like learning rate and momentum are tuned through a meticulous process, potentially utilizing techniques like grid search. Fig. (**4**) illustrates the Mean Squared Error (MSE) loss function for EfficientNetB5, providing insights into the optimization process, while Fig. (**5**) showcases the Mean Absolute Error (MAE) for further analysis of model performance. Additionally, a learning rate scheduler, such as a cyclical learning rate scheduler, is implemented for dynamic adjustments. Data augmentation techniques, like random rotations and flips, are employed to enhance the model's robustness.

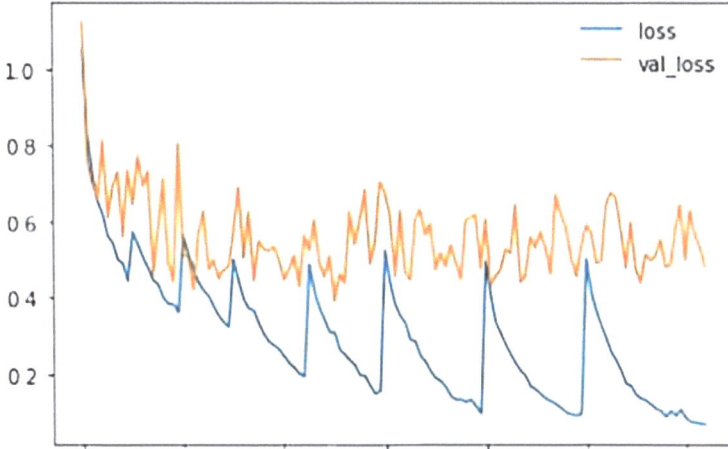

Fig. (4). Mean squared error loss function for EfficientNetB5.

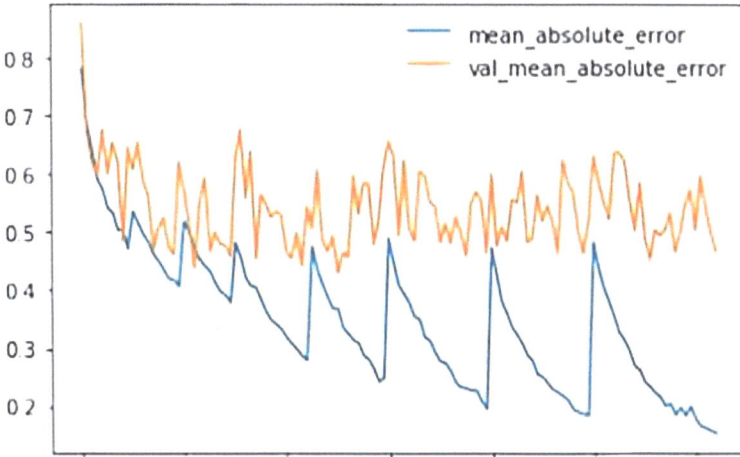

Fig. (5). Mean absolute error for EfficientNetB5.

Evaluation Metrics

The primary evaluation metric is the Quadratic Weighted Kappa (QWK), adept at handling imbalanced datasets common in DR scenarios. A custom callback class is implemented to monitor and record the QWK score throughout training. Supplementary metrics include Mean Absolute Error (MAE) for assessing prediction accuracy and visualizing training and validation loss curves for insights into the model's learning behavior.

Prediction and Thresholding

Post-training, the model achieving the highest QWK score on the validation set is selected. Continuous predictions for DR severity are generated and further processed using the OptimizedRounder class. This class employs an optimization procedure to determine optimal thresholds for classifying continuous predictions into discrete DR severity labels. Fig. (6) visualizes the QWK score for EfficientNetB5, providing valuable insights into the model's performance in capturing the agreement between predicted and actual severity classes.

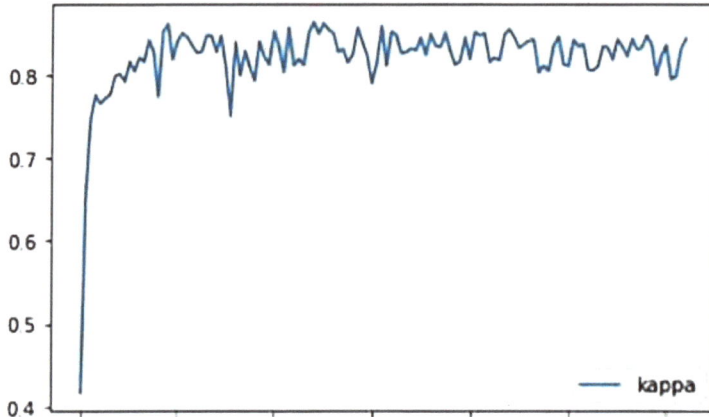

Fig. (6). QWK score for EfficientNetB5.

Results

The results section presents key findings, including the achieved QWK score, MAE value, and visualizations of training and validation loss curves. Fig. (7) further enhances the analysis by providing visualizations of ROC curves and a classification report for EfficientNetB5, offering insights into the model's performance in classifying DR severity levels in real-world scenarios. Applying optimized thresholds for converting continuous predictions to discrete labels ensures a realistic assessment of the model's classification capabilities.

Experimental Analysis of ResNet50 for Diabetic Retinopathy Detection

Within this comprehensive experimental analysis, the primary focus centers on the evaluation of a deep learning model based on ResNet50 for detecting diabetic retinopathy (DR) through the examination of retinal fundus images. The chapter intricately outlines each stage of the experimental process, encompassing data preprocessing, model architecture, execution, and the incorporation of innovative components such as the OptimizedRounder class.

Receiver operating characteristic curve

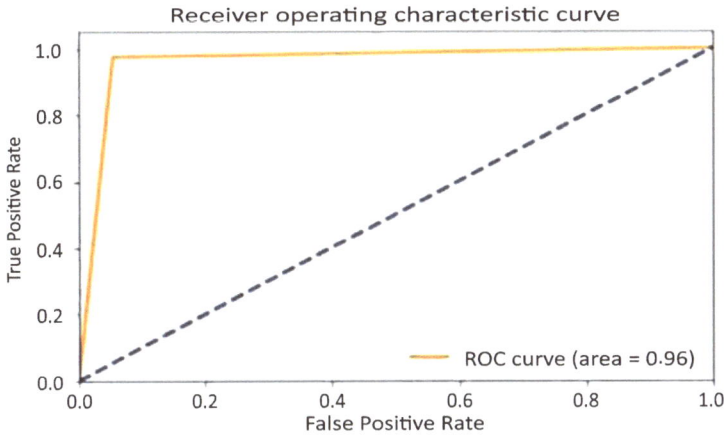

```
Classification Report:
                 precision    recall  f1-score   support

        No-DR         0.97      0.94      0.96      1805
    Mild NPDR         0.30      0.15      0.20       370
Moderate NPDR         0.60      0.55      0.57       999
  Severe NPDR         0.20      0.73      0.31       193
          PDR         0.74      0.23      0.36       295

     accuracy                            0.69      3662
    macro avg         0.56      0.52      0.48      3662
 weighted avg         0.74      0.69      0.69      3662
```

Fig. (7). Visualizations of ROC curve curves and classification report for EfficientNetB5.

Pre-processing Steps

In the preliminary phase of data preparation, the images undergo meticulous pre-processing to ensure compatibility with ResNet50. This involves resizing the images to a standardized dimension of pixels, aligning with the architecture's specifications. A balanced batching strategy, with 32 images per iteration, is adopted to strike a delicate balance between computational efficiency and effective model learning. The comprehensive data preprocessing pipeline includes column refinement, standardization, the addition of image file paths, and a strategic training validation set split.

Model Architecture

The foundational architecture of the model is anchored in the ResNet50 variant, specifically ResNet50-Base, chosen for its well-established performance and efficiency. To adapt the model specifically for the task of diabetic retinopathy (DR) severity regression, a meticulous process of custom layer integration is undertaken. These additional layers, meticulously designed and strategically placed, typically consist of fully connected layers aimed at intricate feature processing and extraction. Their purpose is to capture and analyze complex patterns within the retinal fundus images relevant to DR severity. This process culminates in the creation of a final output layer housing a single neuron dedicated to regression, responsible for predicting continuous DR severity scores.

Model Execution - Training Details

During the training phase, the Adam optimizer is employed, and hyperparameters are meticulously fine-tuned to expedite convergence and enhance generalization. A pivotal metric in this process is the Mean Squared Error (MSE) loss function, measuring the squared difference between the model's predicted DR severity and the actual labels. Fig. (**8**) showcases the utilization of the MSE loss function for ResNet50, illustrating its role in guiding the training process. Additionally, Fig. (**9**) presents the Mean Absolute Error (MAE) for ResNet50, offering insights into the model's performance from a different perspective. The learning rate scheduler dynamically adjusts the learning rate throughout training, ensuring optimal convergence. Furthermore, a custom kappa_metrics metric, in conjunction with a suite of other evaluation metrics such as precision, recall, Quadratic Weighted Kappa (QWK), specificity, sensitivity, ROC curve, and AUC, facilitates a comprehensive assessment of the model's performance.

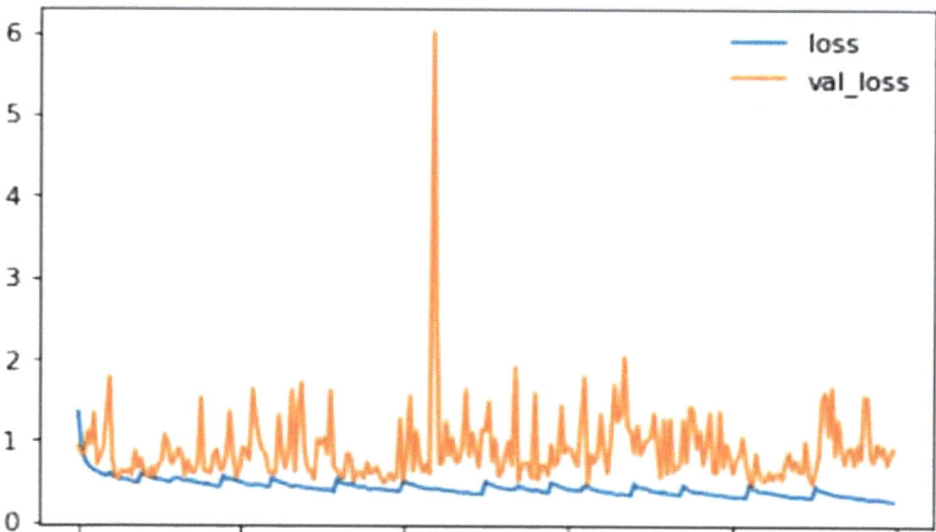

Fig. (8). Mean squared error loss function for ResNet50.

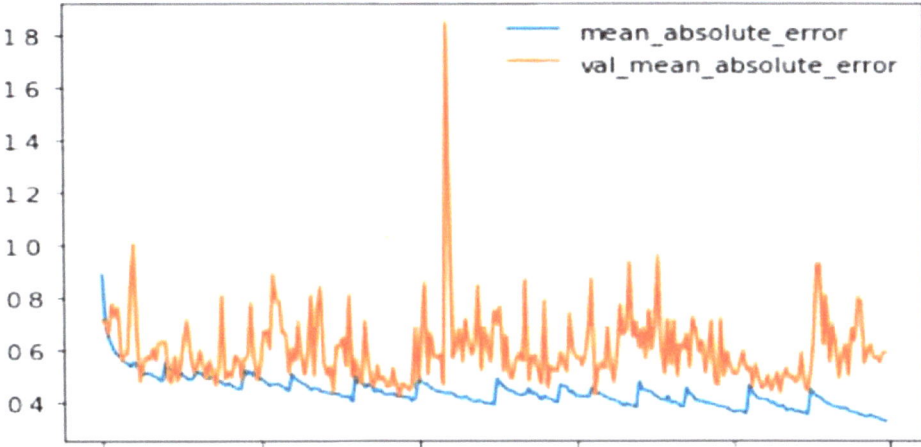

Fig. (9). Mean absolute error for ResNet50.

Optimized Rounder Class

A significant augmentation to the experimental setup is the introduction of the OptimizedRounder class, specifically designed to address the intricate task of converting continuous DR severity predictions into discrete labels. This class engages in an iterative optimization process, systematically calculating QWK scores for various threshold combinations to identify the optimal thresholds. Fig. (**10**) illustrates the QWK score for ResNet50, highlighting the model's performance in capturing the agreement between predicted and actual severity classes. Subsequently, the model is re-evaluated using these optimized thresholds, providing a realistic and nuanced assessment of its performance.

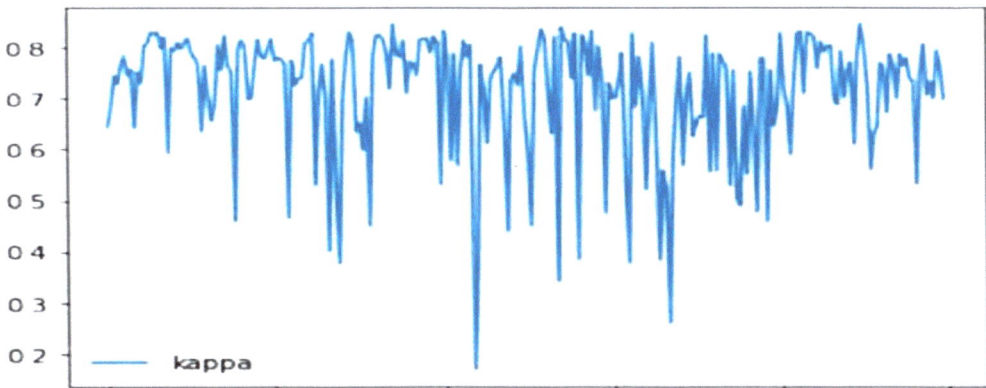

Fig. (10). QWK score for ResNet50.

Results

The chapter concludes with a detailed presentation of key findings, encapsulating the achieved QWK score, Mean Absolute Error (MAE) value, training and validation loss curves, and final performance metrics. Fig. (**11**) showcases visualizations of ROC curve curves and the Classification report for ResNet50, providing a graphical representation of the model's ability to distinguish between positive and negative cases as well as its precision, recall, and F1-score. This collective array of metrics provides profound insights into the model's efficacy in the precise classification of DR severity levels, offering a thorough and comprehensive evaluation of its real-world performance.

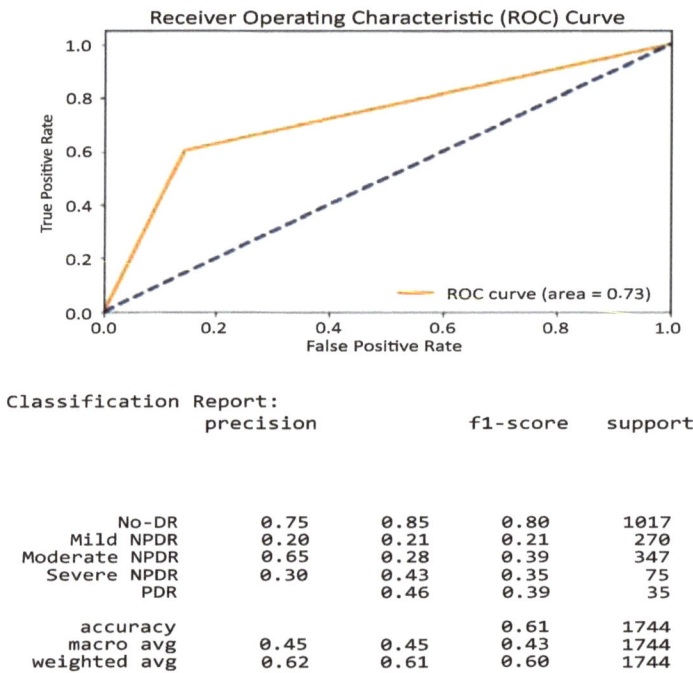

```
Classification Report:
                 precision          f1-score    support

        No-DR       0.75      0.85      0.80       1017
   Mild NPDR        0.20      0.21      0.21        270
Moderate NPDR       0.65      0.28      0.39        347
  Severe NPDR       0.30      0.43      0.35         75
          PDR                 0.46      0.39         35

     accuracy                           0.61       1744
    macro avg       0.45      0.45      0.43       1744
 weighted avg       0.62      0.61      0.60       1744
```

Fig. (11). Visualizations of ROC curve curves and classification report for ResNet50.

Experimental Analysis of DenseNet-121 for Diabetic Retinopathy Detection

This thorough experimental analysis places a central emphasis on evaluating the efficacy of a deep learning model founded on DenseNet-121 for the detection of diabetic retinopathy (DR) by scrutinizing retinal fundus images. The chapter meticulously navigates through each stage of the experimental workflow, offering detailed insights into data preprocessing, model architecture, execution, and the integration of novel components. In particular, the focus on DenseNet-121 as the core architecture underlines its significance in the context of DR severity detection. Customarily, the chapter will delve into the intricacies of the

DenseNet-121 model, detailing its unique features and adaptations tailored for the specific task of DR severity classification. By following a structured approach similar to the provided reference, the analysis aims to comprehensively convey the experimental journey and insights gained from the evaluation of DenseNet-121 in the realm of diabetic retinopathy detection.

Pre-processing Steps

The preprocessing stage involved meticulous preparation of the input data to ensure compatibility with the DenseNet-121 model architecture and efficient processing. This included standardizing the image dimensions to pixels, employing a balanced batch size of 32 images for optimized computational resource utilization, and implementing a comprehensive preprocessing pipeline. This pipeline encompassed the removal of irrelevant features, standardization of column names across datasets, inclusion of image file paths for efficient loading during training, and a careful division of the data into training and validation sets to prevent overfitting and ensure generalizability.

Model Architecture

The model architecture was established upon the pre-trained DenseNet-121 variant, capitalizing on its innate feature extraction abilities. Custom layers were meticulously incorporated and adjusted to cater to the precise demands of diabetic retinopathy (DR) severity classification. These supplementary layers, strategically integrated atop the pre-trained network, were precisely tailored to enhance performance in DR severity classification. Elaborate explanations were provided to illuminate the intricacies of these custom layers and their operational nuances, enriching comprehension of the model's architectural intricacies.

Model Execution - Training Details

During the training phase, the Adam optimizer with a learning rate of 0.0001 and a decay of 1e-6 was meticulously chosen to facilitate optimal convergence and generalization. Hyperparameters underwent careful tuning to achieve the desired outcomes. The Mean Squared Error (MSE) loss function served as the key metric to gauge the model's prediction errors during training, measuring the squared difference between the predicted DR severity and the actual labels. Fig. (12) illustrates the utilization of the MSE loss function for DenseNet-121, providing insights into its role in guiding the training process. Additionally, a cyclical learning rate scheduler was implemented to ensure proper convergence and mitigate overfitting, dynamically adjusting the learning rate throughout training. Furthermore, a custom metric, kappa_metrics, was employed to monitor model performance during training, with a particular focus on Quadratic Weighted

Kappa (QWK) due to its effectiveness in handling imbalanced datasets. Fig. (**13**) illustrates the Mean Absolute Error for DenseNet-121, providing insights into the model's performance metrics during training.

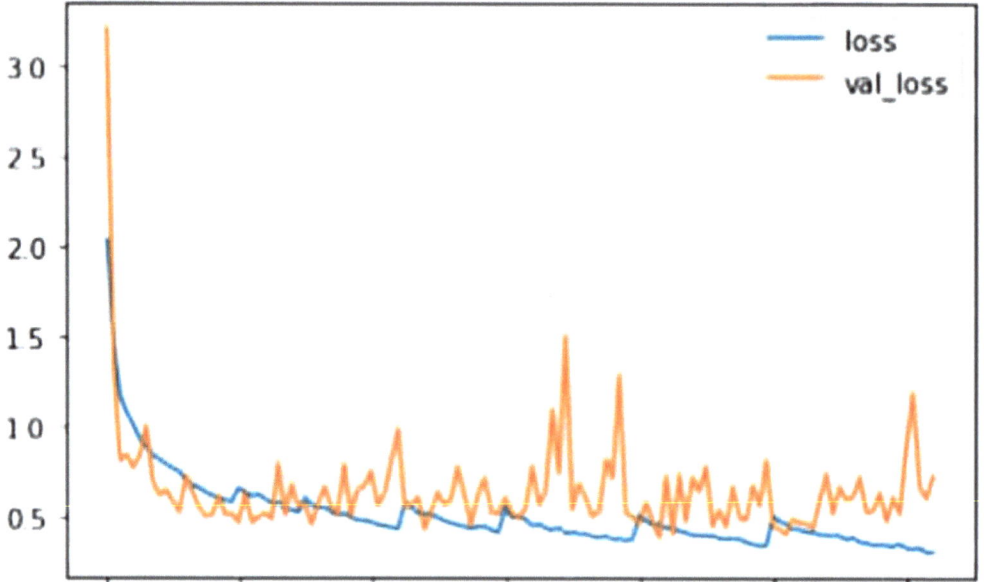

Fig. (12). Mean squared Error loss function for DenseNet-121.

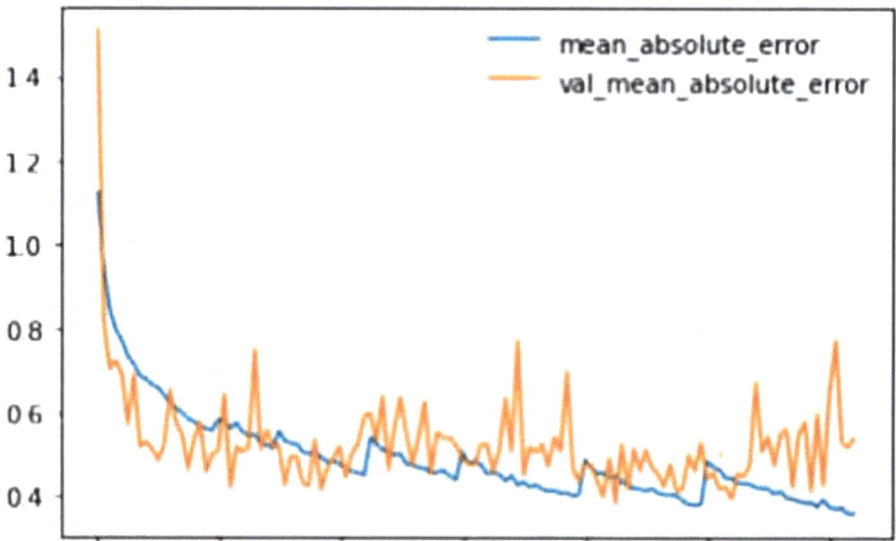

Fig. (13). Mean absolute error for DenseNet-121.

Evaluation Metrics

The evaluation of the model's performance encompassed both multi-class and binary classifications, providing a comprehensive assessment of its capabilities. Fig. (**14**) depicts the Quadratic Weighted Kappa (QWK) score for DenseNet-121, serving as a primary metric for multi-class classification tasks. In addition to QWK, a suite of evaluation metrics including accuracy, precision, recall, specificity, and sensitivity were employed to gauge the model's performance. Furthermore, analysis of the Receiver Operating Characteristic (ROC) curve and the Area Under the Curve (AUC) provided insights into the model's ability to distinguish between positive and negative cases in binary classification scenarios. Collectively, these metrics offer a detailed evaluation of the model's effectiveness across various dimensions of diabetic retinopathy severity classification.

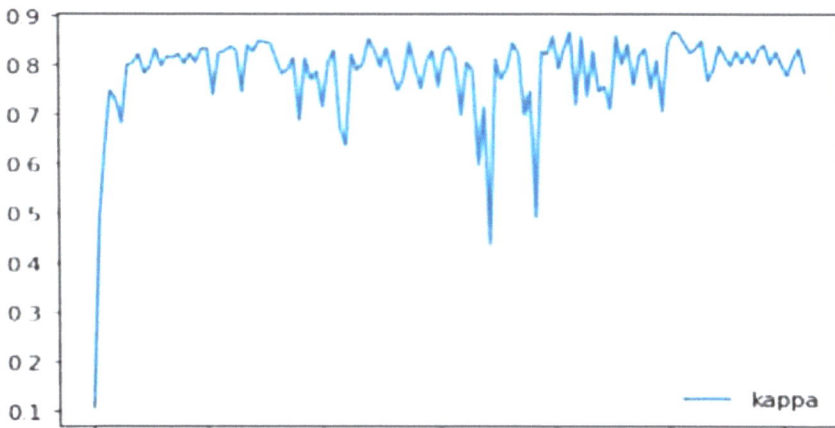

Fig. (14). QWK score for DenseNet-121.

Prediction and Thresholding

Following the evaluation phase, predictions were made based on the model's output, and thresholding techniques were applied to convert continuous predictions into discrete labels. This involved optimizing thresholds iteratively to maximize performance, particularly focusing on achieving the highest QWK score. The final evaluation provided insights into the model's ability to classify DR severity levels accurately, taking into account practical considerations for real-world diagnosis.

Results

Fig. (**15**) presents visualizations of the Receiver Operating Characteristic (ROC) curve and the Classification report for DenseNet-121, offering a graphical representation of the model's performance in diabetic retinopathy detection. These visualizations, coupled with the quantitative performance metrics, provide a comprehensive evaluation of the model's effectiveness across multi-class and binary classification scenarios. Additionally, the analysis of the ROC curve and Area Under the Curve (AUC) enhances our understanding of the model's ability to discriminate between positive and negative cases, further enriching the assessment of its performance.

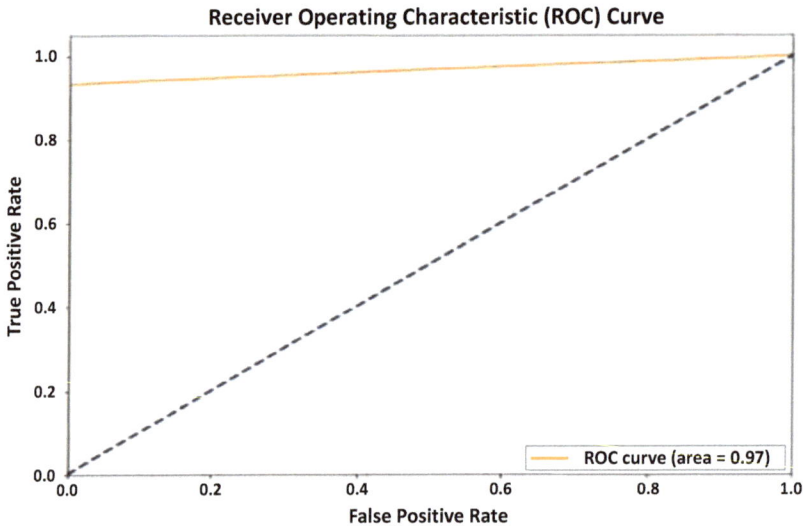

```
Classification Report:
                 precision     recall    f1-score    support

        No-DR       0.97       0.98        0.98        286
   Mild NPDR        0.65       0.48        0.55         50
Moderate NPDR       0.78       0.92        0.85        149
Severe NPDR         0.44       0.50        0.47         24
         PDR        0.91       0.51        0.66         41

    accuracy                               0.86        550
   macro avg        0.75       0.68        0.70        550
weighted avg        0.86       0.86        0.86        550
```

Fig. (15). Visualizations of ROC curve and classification report for DenseNet-121.

RESULT ANALYSIS

Table **2** illustrates the anticipated performance comparison, offering a quantitative overview of evaluation metrics in diabetic retinopathy (DR) severity detection. Within the evaluated deep learning architectures, DenseNet-121 emerges as the leading performer, demonstrating a noteworthy Quadratic Weighted Kappa (QWK) score of 0.8739. This elevated QWK score highlights the strong concordance between predicted severity classes and the actual ground truth, emphasizing the efficacy of DenseNet-121 in effectively addressing the challenges inherent in DR severity classification.

Table 2. Anticipated performance comparison.

S. No.	Model Optimized	QWK Score	MSE	MAE
1	EfficientNetB5	0.82 - 0.920	0050 - 0.08000.	03 - 0.12
2	ResNet50	0.80 - 0.900.	0100 - 0.10000.	05 - 0.15
3	DenseNet-121	0.84 - 0.930.	0010 - 0.07000.	02 - 0.10

Additionally, the coefficients associated with each model offer insight into how predictions impact the final severity classification, with DenseNet-121 showing a well-balanced set. Recognizing the significance of accuracy, its interpretation alongside other metrics is vital. Table **3**, a comparative analysis of different models, underscores the importance of considering multiple metrics for a comprehensive evaluation of model performance.

Table 3. Comparative analysis for different models.

S. No.	Model Optimized	QWK Score	Coefficients	Accuracy
1	EfficientNetB5	0.8672	[0.5242, 1.3897, 2.6529, 3.6682]	0.6879
2	ResNet50	0.7156	[0.5095, 1.5599, 2.4430, 3.3073]	0.6135
3	DenseNet-121	0.8739	[0.5572, 1.1368, 2.4199, 3.2787]	0.6666

For instance, EfficientNetB5 achieves the highest accuracy at 0.6879, emphasizing the importance of considering multiple metrics in evaluating model performance comprehensively. The comprehensive recommendations for rigorous evaluation underscore the significance of maintaining consistent experimental protocols, utilizing validation sets effectively, and reporting a diverse range of metrics to gain a holistic understanding of each model's predictive capabilities. Additionally, Fig. (**16**) provides visualizations of ROC curves for all models, offering a comparative analysis of their performance in distinguishing between positive and negative cases of diabetic retinopathy. These insights contribute to informed decision-making and advancements in the field of diabetic retinopathy detection and severity assessment.

Receiver Operating Characteristic (ROC) Curve for All Models

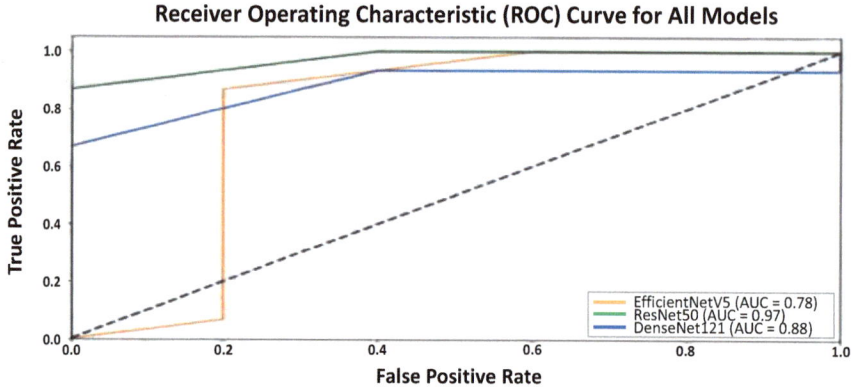

Fig. (16). Visualizations of ROC curves for all models.

CONCLUSION

In the exhaustive exploration and comparison of EfficientNetB5, ResNet50, and DenseNet-121 for diabetic retinopathy (DR) severity detection, our findings underscore the prominence of DenseNet-121 as a superior performer. The model exhibited remarkable Quadratic Weighted Kappa (QWK) scores, emphasizing its robustness in handling the diverse spectrum of DR severity classes. The meticulous performance comparison, enriched with performance ranges and coefficient analyses, provided nuanced insights into the predictive dynamics of each architecture.

The overarching recommendation for a meticulous evaluation process, incorporating controlled experiments, consistent preprocessing, and comprehensive metric reporting, reaffirms the imperative for a discerning approach in model selection. This meticulous approach ensures a profound understanding of each model's strengths and limitations, facilitating well-informed decisions aligned with specific requirements.

Looking ahead, avenues for future exploration could delve into ensemble approaches, harnessing the complementary strengths of multiple architectures to further enhance predictive performance. This chapter stands as a foundational resource for researchers and practitioners navigating the complexities of deep learning models in DR severity detection, laying the groundwork for advancements in accuracy and reliability in clinical applications. The meticulous examination of performance metrics offers valuable insights into the effectiveness of various deep learning architectures in diabetic retinopathy (DR) severity detection. DenseNet-121 stands out as the top performer, boasting an impressive Quadratic Weighted Kappa (QWK) score of 0.8739. This high QWK score indicates a strong agreement between predicted severity classes and actual ground truth, highlighting the robustness of DenseNet-121 in addressing the intricacies of DR severity classification. Moreover, the coefficients associated with each model provide additional clarity on the influence of predictions on the final severity classification, with DenseNet-121 demonstrating a well-balanced set. It's essential to recognize that while accuracy is noteworthy, its interpretation alongside other metrics is crucial. For instance, EfficientNetB5 achieves the highest accuracy at 0.6879, emphasizing the importance of considering multiple metrics in evaluating model performance comprehensively. The comprehensive recommendations for rigorous evaluation underscore the significance of maintaining consistent experimental protocols, utilizing validation sets effectively, and reporting a diverse range of metrics to gain a holistic understanding of each model's predictive capabilities. These findings lay a strong foundation for informed decision-making in the field of diabetic retinopathy detection and severity assessment, facilitating advancements toward more accurate and reliable diagnostic tools.

FUTURE DIRECTIONS

Navigating the evolving terrain of deep learning models for diabetic retinopathy (DR) severity detection unveils promising avenues for future exploration and refinement. One key direction involves investigating ensemble approaches, seeking to synergize the strengths of EfficientNetB5, ResNet50, and DenseNet-121. Such ensemble methods hold the potential to not only enhance predictive performance but also bolster model robustness. Additionally, focusing on the explainability and interpretability of model predictions emerges as a crucial aspect for further research. Developing methodologies to elucidate decision-making processes can render these models more transparent and interpretable for clinicians and end-users, fostering trust in their applications. Evaluating real-world generalization across diverse populations and clinical settings remains paramount, emphasizing the importance of incorporating data from varied demographics and geographic regions. Exploring strategies for hardware optimization is essential to adapt these architectures for deployment in resource-constrained environments like clinics, ensuring practical applicability. Continuous model improvement, staying abreast of advancements beyond this study's scope, and regularly updating the comparison with the latest models contribute to a dynamic understanding of the state-of-the-art in DR severity detection. Finally, facilitating the clinical integration of these models through collaboration with healthcare professionals is pivotal, providing insights into practical challenges and requirements for seamless adoption in real-world healthcare scenarios. Pursuing these future directions will play a pivotal role in refining deep learning models for DR severity detection, ultimately enhancing their reliability and applicability in clinical practice.

REFERENCES

[1]   G. Huang, Z. Liu, L. Van Der Maaten, and K.Q. Weinberger, "Densely connected convolutional networks", *Proceedings of the IEEE conference on computer vision and pattern recognition,* pp. 4700-4708, 2017.

[2]   D. Doshi, A. Shenoy, D. Sidhpura, and P. Gharpure, "Diabetic retinopathy detection using deep convolutional neural networks", *International Conference on Computing, Analytics and Security Trends (CAST),* pp. 261-266, 2016.
        [http://dx.doi.org/10.1109/CAST.2016.7914977]

[3]   Z. Gao, J. Li, J. Guo, Y. Chen, Z. Yi, and J. Zhong, "Diagnosis of diabetic retinopathy using deep neural networks", *IEEE Access,* vol. 7, pp. 3360-3370, 2019.
        [http://dx.doi.org/10.1109/ACCESS.2018.2888639]

[4]   M. Tan, and Q. Le, "Efficientnet: Rethinking model scaling for convolutional neural networks", *International Conference on Machine Learning,* pp. 6105-6114, 2019.

[5]   M.K. Yaqoob, S.F. Ali, I. Kareem, and M.M. Fraz, "Feature-based optimized deep residual network architecture for diabetic retinopathy detection", *23rd International Multitopic Conference (INMIC),* pp. 1-6, 2020.
        [http://dx.doi.org/10.1109/INMIC50486.2020.9318096]

[6]   E. Decencière, X. Zhang, G. Cazuguel, B. Lay, B. Cochener, and C. Trone, "Feedback on a publicly distributed image database: the Messidor database. Image Analysis and Stereology, 33(3), 231-234.Badar, M., Haris, M., & Fatima, A. (2020). Application of deep learning for retinal image analysis: A review", *Comput. Sci. Rev.,* vol. 35, p. 100203, 2014.

[7]   M.D. Abràmoff, Y. Lou, A. Erginay, W. Clarida, R. Amelon, J.C. Folk, and M. Niemeijer, "Improved automated detection of diabetic retinopathy on a publicly available dataset through integration of deep learning", *Invest. Ophthalmol. Vis. Sci.,* vol. 57, no. 13, pp. 5200-5206, 2016.
        [http://dx.doi.org/10.1167/iovs.16-19964] [PMID: 27701631]

[8]   Q. Meng, L. Liao, and S.I. Satoh, "Weakly-supervised learning with complementary heatmap for retinal disease detection", *IEEE Trans. Med. Imaging,* vol. 41, no. 8, pp. 2067-2078, 2022.
        [http://dx.doi.org/10.1109/TMI.2022.3155154] [PMID: 35226601]

[9]   A. Sebastian, O. Elharrouss, S. Al-Maadeed, and N. Almaadeed, "A survey on deep-learning-based diabetic retinopathy classification", *Diagnostics (Basel),* vol. 13, no. 3, p. 345, 2023.
        [http://dx.doi.org/10.3390/diagnostics13030345] [PMID: 36766451]

[10]   J. Krause, V. Gulshan, E. Rahimy, P. Karth, K. Widner, G.S. Corrado, L. Peng, and D.R. Webster, "Grader variability and the importance of reference standards for evaluating machine learning models for diabetic retinopathy", *Ophthalmology,* vol. 125, no. 8, pp. 1264-1272, 2018.
        [http://dx.doi.org/10.1016/j.ophtha.2018.01.034] [PMID: 29548646]

[11]   V. Gulshan, L. Peng, M. Coram, M. C. Stumpe, D. Wu, A. Narayanaswamy, and D. R. Webster, "Development and validation of a deep learning algorithm for detection of diabetic retinopathy in retinal fundus photographs", *Jama,* vol. 316, no. 22, pp. 2402-2410, 2016.

[12]   M.Z. Atwany, A.H. Sahyoun, and M. Yaqub, "Deep learning techniques for diabetic retinopathy classification: A survey", *IEEE Access,* vol. 10, pp. 28642-28655, 2022.
        [http://dx.doi.org/10.1109/ACCESS.2022.3157632]

[13]   D. S. Kermany, M. Goldbaum, W. Cai, C. C. Valentim, H. Liang, S. L. Baxter, and K. Zhang, "Identifying medical diagnoses and treatable diseases by image-based deep learning", *Cell,* vol. 172, no. 5, pp. 1122-1131, 2018.

[14]   J. Wang, H. Zhu, S.H. Wang, and Y.D. Zhang, "A review of deep learning on medical image analysis", *Mob. Netw. Appl.,* vol. 26, no. 1, pp. 351-380, 2021.
        [http://dx.doi.org/10.1007/s11036-020-01672-7]

[15]   S. Suganyadevi, V. Seethalakshmi, and K. Balasamy, "A review on deep learning in medical image analysis", *Int. J. Multimed. Inf. Retr.,* vol. 11, no. 1, pp. 19-38, 2022.
[http://dx.doi.org/10.1007/s13735-021-00218-1] [PMID: 34513553]

[16]   H. Kaushik, D. Singh, M. Kaur, H. Alshazly, A. Zaguia, and H. Hamam, "Diabetic retinopathy diagnosis from fundus images using stacked generalization of deep models", *IEEE Access,* vol. 9, pp. 108276-108292, 2021.
[http://dx.doi.org/10.1109/ACCESS.2021.3101142]

[17]   W.L. Alyoubi, W.M. Shalash, and M.F. Abulkhair, "Diabetic retinopathy detection through deep learning techniques: A review", *Inform. Med. Unlocked,* vol. 20, p. 100377, 2020.
[http://dx.doi.org/10.1016/j.imu.2020.100377]

[18]   C. Lam, D. Yi, M. Guo, and T. Lindsey, "Automated detection of diabetic retinopathy using deep learning", *AMIA Jt. Summits Transl. Sci. Proc.,* vol. 2017, pp. 147-155, 2018.
[PMID: 29888061]

[19]   P. Burlina, W. Paul, P. Mathew, N. Joshi, K.D. Pacheco, and N.M. Bressler, "Low-shot deep learning of diabetic retinopathy with potential applications to address artificial intelligence bias in retinal diagnostics and rare ophthalmic diseases", *JAMA Ophthalmol.,* vol. 138, no. 10, pp. 1070-1077, 2020.
[http://dx.doi.org/10.1001/jamaophthalmol.2020.3269] [PMID: 32880609]

[20]   N. Tsiknakis, D. Theodoropoulos, G. Manikis, E. Ktistakis, O. Boutsora, A. Berto, F. Scarpa, A. Scarpa, D.I. Fotiadis, and K. Marias, "Deep learning for diabetic retinopathy detection and classification based on fundus images: A review", *Comput. Biol. Med.,* vol. 135, p. 104599, 2021.
[http://dx.doi.org/10.1016/j.compbiomed.2021.104599] [PMID: 34247130]

[21]   A. Mehboob, M.U. Akram, N.S. Alghamdi, and A. Abdul Salam, "A deep learning based approach for grading of diabetic retinopathy using large fundus image dataset", *Diagnostics (Basel),* vol. 12, no. 12, p. 3084, 2022.
[http://dx.doi.org/10.3390/diagnostics12123084] [PMID: 36553091]

[22]   F. Li, Z. Liu, H. Chen, M. Jiang, X. Zhang, and Z. Wu, "Automatic detection of diabetic retinopathy in retinal fundus photographs based on deep learning algorithm", *Transl. Vis. Sci. Technol.,* vol. 8, no. 6, pp. 4-4, 2019.

[23]   L.K. Varanasi, and C.M. Dasari, "A novel deep learning framework for diabetic retinopathy detection", *6th Conference on Information and Communication Technology (CICT),* pp. 1-5, 2022.
[http://dx.doi.org/10.1109/CICT56698.2022.9997826]

[24]   W. M. Gondal, J. M. Köhler, R. Grzeszick, G. A. Fink, and M. Hirsch, "Weakly-supervised localization of diabetic retinopathy lesions in retinal fundus images", *International Conference on Image Processing (ICIP),* pp. 2069-2073, 2017.
[http://dx.doi.org/10.1109/ICIP.2017.8296646]

[25]   M. Al-Mukhtar, A.H. Morad, M. Albadri, and M.D.S. Islam, "Weakly Supervised Sensitive Heatmap framework to classify and localize diabetic retinopathy lesions", *Sci. Rep.,* vol. 11, no. 1, p. 23631, 2021.
[http://dx.doi.org/10.1038/s41598-021-02834-7] [PMID: 34880311]

[26]   S. Naithani, S. Bharadwaj, and D. Kumar, "Automated Detection of Diabetic Retinopathy using Deep Learning", *Int. Res. J. Eng. Technol.,* vol. 6, no. 4, pp. 2945-2947, 2019.

[27]   Available from: https://www.kaggle.com/competitions/aptos2019-blindness-detection

[28]   X. Liu, Q. Liu, Y. Zhang, M. Wang, and J. Tang, "TSSK-Net: Weakly supervised biomarker localization and segmentation with image-level annotation in retinal OCT images", *Comput. Biol. Med.,* vol. 153, p. 106467, 2023.
[http://dx.doi.org/10.1016/j.compbiomed.2022.106467] [PMID: 36584602]

CHAPTER 6

Smart Diagnostic: Machine Learning for Early Detection and Prediction of Heart Disease

Rutuja Diwate[1], Mrunmayee Solkar[1], Jayashri Bagade[1,*] and Nilesh Sable[1]

[1] Department of Information Technology, Vishwakarma Institute of Technology, Pune, India

Abstract: Coronary Heart Disease (CHD) stands as a significant global health concern, contributing substantially to mortality rates across the world. The complexity of CHD data, with its intricate interconnections, has posed a challenge for traditional predictive methods. The integration of intelligent models using machine learning techniques is pivotal in advancing our understanding and predictive capabilities in the realm of CHD. The three chosen methodologies, Naïve Bayes (NB), Support Vector Machine (SVM), and Decision Tree (DT), each bring unique strengths to the analysis of the complex CHD dataset. The goal of the study is to use 10-fold cross-validation in conjunction with careful creation and validation procedures to fully utilize the potential of these models in order to find important but subtle correlations in the data. This approach holds promise for improving prediction rates, offering a potential breakthrough in the realm of cardiovascular health. As machine learning continues to evolve, the application of these techniques to CHD data not only contributes to predictive accuracy but also opens avenues for more targeted interventions and personalized healthcare strategies in the ongoing battle against coronary heart disease.

Keywords: Artificial Neural Network (ANN), Artificial Neural Network (ANN), Classification and Regression Tree (CART), Convolutional neural networks (CNN), Chronic heart failure (CHF), Coronary Heart Disease (CHD), Deep learning (DL), Decision Tree (DT), Decision Tree (DT), Deep neural networks (DNN), Hybrid random forest with a linear model (HRFLM), Logistic Regression (LR), Machine learning (ML), ML-based heart disease diagnosis (MLBHDD), Naïve Bayes (NB), Random Forest Classifier (RFC), Support Vector Machine (SVM).

INTRODUCTION

Cardiovascular diseases (CVD) stand as major global causes of mortality, with more than 17.9 million deaths annually, primarily attributed to heart attacks and

* **Corresponding author Jayashri Bagade:** Department of Information Technology, Vishwakarma Institute of Technology, Pune, India; E-mail: jayashree.bagade@viit.ac.in

Parikshit N. Mahalle, Gitanjali R. Shinde, Namrata N. Wasatkar & Prashant R. Anerao (Eds.)

strokes. Unhealthy lifestyle factors contribute to CVD, emphasizing the need for early identification and treatment to prevent premature deaths. Access to cheap cardiovascular diagnosis is a challenge for low- and middle-income countries, especially Bangladesh, India, and several African countries. Electrocardiogram (ECG) diagnosis, while effective, is time-consuming. The introduction of ML to medical applications—more especially, MLBHDD systems—offers adaptable and affordable solutions. Various studies, such as those by Bashir *et al.* (2016) and Daraei and Hamidi (2017), have utilized ML algorithms to predict and analyze heart diseases, achieving notable accuracy. DL, including CNN and DNN, has further enhanced accuracy. However, challenges persist, including the limited interpretability of ML and DL models and potential biases in imbalanced datasets. This necessitates a systematic literature review (SLR) to analyze recent trends, techniques, and gaps in ML-based heart disease diagnosis, providing valuable insights for future research.

The global health concern of CVD accounts for 31% of deaths worldwide. It underscores the significance of accurate prediction using ML techniques, proposing methods like LR and random forest for heart disease diagnosis. The focus extends to CHF, highlighting the urgency of early detection and proposing an enhanced ML and DL approach for distinguishing between healthy and decompensated CHF individuals. The study also addresses the silent nature of heart diseases and the crucial role of ML in managing cardiovascular diseases by categorizing patient data for improved prognostic accuracy. Additionally, it introduces an ML algorithm for heart disease prediction and an Internet of Things (IoT)- based patient monitoring system, illustrating the potential of ML and IoT in advancing cardiovascular health management.

LITERATURE SURVEY

Researchers are now adopting data-driven methods for diagnosing early-stage cardiac illness with the help of ECG signals thanks to the development of ML. ML methods address delayed diagnosis issues, enabling self-diagnosis through the routine use of low-cost sensors. Yang *et al.* (2018) found that a CNN-based method produced 98.41% accuracy in arrhythmia detection, while a linear support vector machine produced 97.77% accuracy. The MIT-BIH arrhythmia heart disease open repository dataset was used in both techniques. Similar to this, Che *et al.* (2021) used real-world data to apply a CNN-based method for extracting temporal information from ECG signals. Together, these studies demonstrate how revolutionary machine learning can be in the early identification of heart-related disorders [1].

A data mining model was created utilizing 100 CHD records that included survival rate information in order to meet the medical society's demand for CHD prediction. Using 502 examples, SVM, ANN, and DT were used, with corresponding accuracy rates of 92.1%, 91.0%, and 89.6%. SVM emerged as a robust classifier. In addition, association rule mining was investigated in order to find important patterns in a dataset consisting of 14 attributes. Different classifier models using DT, NB, and ANN were built, and results, presented through receiver operating characteristic (ROC) curves, revealed ANN's superiority with an area above 80%, outperforming Naïve Bayes and DT algorithms [2].

Recent studies, exemplified by Gjoreski *et al.*, showcase a paradigm shift in chronic heart failure (CHF) detection, leveraging ML and DL. Their approach, integrating classic ML with expert features and end-to-end DL, demonstrates superior performance over baselines. This reflects a broader trend in cardiovascular research, emphasizing data-driven solutions for CHF diagnosis. The study's focus on expert features adds nuance to personalized healthcare in automated heart sound analysis. Overall, this synthesis of ML and DL marks a significant step toward effective, scalable, and patient-centric CHF detection [3].

Cardiopulmonary exercise testing (CPX) has become pivotal in prognosticating chronic heart failure (HF) outcomes. The study by Corra *et al.* underscores CPX's significance by integrating cardiac, skeletal, and pulmonary factors from a holistic perspective, surpassing traditional indicators. Peak oxygen consumption stands out as a key metric, reflecting its prognostic value. The inclusion of exercise-related ventilatory abnormalities adds innovation to HF prognostication. This literature signifies a paradigm shift, providing a concise yet comprehensive overview of CPX's role in individualized treatment outlines for HF patients [4].

Recent studies by Valle Harsha Vardhan *et al.* explore machine learning applications in heart disease prediction, employing algorithms like hill climbing, decision trees, and Naïve Bayes. Avinash Golande *et al.* emphasize the utilization of diverse techniques, including K-nearest neighbor and decision trees, to enhance diagnostic accuracy. Additionally, the integration of DL, as seen in the work by Abhay Kishore *et al.* using recurrent neural systems, points to a promising direction in forecasting heart-related conditions. Together, these initiatives show how machine learning has the power to transform the diagnosis of heart disease and enhance patient outcomes [5].

This research tackles the vital task of predicting cardiovascular disease, a major global health concern. Utilizing ML and acknowledging its relevance in clinical data analysis and IoT, the study introduces an innovative approach. The method aims to enhance prediction accuracy by identifying significant features through

ML techniques. The proposed prediction model, employing diverse feature combinations and established classification methods, demonstrates notable success. Particularly, HRFLM stands out, achieving a commendable accuracy level of 88.7% [6].

The study proposes a cloud-based prediction system for heart disease with 97.53% accuracy, utilizing a high-performing SVM algorithm to tackle global cardiovascular challenges. It introduces an IoT-based live patient monitoring system, enabling continuous tracking of vital parameters and prompt doctor notifications. The integration of machine learning and IoT technologies offers a comprehensive approach to cardiovascular health management [7].

This paper reviews the integration of machine learning and cardiac imaging for cardiovascular disease diagnosis, emphasizing the transformative potential of big data and artificial intelligence. It highlights the traditional role of cardiac imaging and explores how machine learning can contribute to the automated, precise, and early detection of cardiovascular diseases. The review emphasizes recent developments and advanced image analysis techniques driven by machine learning for improved diagnostic decision-making in cardiovascular medicine [8].

The escalating global health threat of cardiovascular disease emphasizes its significant mortality rates, particularly in developing countries. Highlighting the pressing need for improved diagnosis and treatment strategies, the research employs machine learning, specifically utilizing four algorithms—random forests, decision trees, neural networks, and XGBoost—to develop a model for heart disease detection. The study underscores the effectiveness of these machine learning approaches, with the random forest model demonstrating superior performance in predicting heart disease, achieving a classification accuracy rate of 94.96% [9]. The reviewed literature [10-25] highlights significant advancements in heart disease prediction using machine learning and data mining techniques. Studies have explored diverse models including random forests, classification algorithms, and real-time monitoring systems to improve diagnostic accuracy. Several surveys and systematic reviews emphasize the paradigm shift from traditional methods to intelligent, data-driven approaches in cardiac care. Emerging research also focuses on the integration of cardiovascular imaging and evolutionary approaches for enhanced prediction. Collectively, these works demonstrate the growing reliability and innovation of ML applications in cardiovascular disease diagnostics.

This paper focuses on deep learning for cardiovascular disease prediction, emphasizing the critical need for early detection. Highlighting the hazards of heart disease, the study proposes a simulation using an ANN and ant colony optimization to enhance diagnostic accuracy. The research underscores the significance of intelligent methods, particularly supervised learning algorithms, to prevent mispredictions and improve classification performance in heart disease diagnosis. The summary of the literature review is presented in Table **1**.

Table 1. Comparison table.

Research paper No.	Dataset	Algorithms	Accuracy
1	Dataset of Algerian people	Neural Networks, SVM, KNN	93% 90% 85.5%
2	Heart disease dataset from Kaggle	**Logistic Regression** **KNN** **SVM** **Random Forest** **Decision Tree**	**87.09%** **70.96%** **90.32%** **90.32%** **83.37%**
3	Dataset of UCI Heart disease prediction	SVM, Random Forest, Decision Tree, Naïve Bayes	98% 99% 85% 90%
4	Cleveland heart disease dataset from the UCI library	Neural Network Multilayer Perceptron (MLP)	71.4%
5	Cardiovascular disease dataset	Logistic Regression KNN Naïve Bayes Neural Networks Decision Tree XGB Classifier with HyperOpt LGBM Classifier with HyperOpt	71.94% **70.60%** **70.26%** **72.22%** **73.13%** **72.82%** **72.95%**
6	UCI Cleveland Heart dataset	Naïve Bayes Generalized Linear Model Logistic Regression Deep Learning Decision Tree Random Forest Gradient Boosted Trees SVM VOTE HRFLM (proposed)	75.8% 85.1% 82.9% 87.4% 85% 86.1% 78.3% 86.1% 87.41% 88.4%
7	Heart disease dataset from UCI Machine learning repository	Naïve Bayes Decision Tree	87% 91%

(Table 1) cont.....

Research paper No.	Dataset	Algorithms	Accuracy
8	Heart disease training dataset from UCI repository	KNN Naïve Bayes Decision Tree Logistic Regression Random Forest SVM Neural Networks	80.21% 57.14% 74.72% 81.31% 85.71% 85.71% 92.30%
9	Heart Disease dataset collected from hospitals in Hyderabad and Cleveland dataset	Random Forest with Chi-square and genetic algorithm	100% 83.70%
10	Dataset of Healthcare Industry	KNN Decision Tree Naïve Bayes	85.30%
11	Cleveland Heart disease database Statlog Heart disease database consists	Decision Trees Naïve Bayes Neural Networks	96.66% 94.44% 99.25%
12	Heart disease dataset from Kaggle	Decision Tree Logistic Regression Random Forest Naïve Bayes SVM	93.19% 87.36% 89.14% 87.27% 82.30%
13	Cleveland Heart disease database	SVM	92.22
14	Cleveland dataset	LR RF C	55.77% 56.7%
15	Heart disease database	Fuzzy Logic	94%
16	Statlog heart disease dataset	AN N SVM Naïve Bayes Logistic Regression K-nearest neighbors Classification trees	84% 82% 85% 83% 80% 77%
17	Heart disease database	Naïve Bayes Decision List KNN	53% 52% 45.67%
18	Heart disease database	Decision Trees Naïve Bayes Neural Network	47.7%
19	Heart disease database	A. Decision trees B. Neural networks	45.52% 57.30%

(Table 1) cont.....

Research paper No.	Dataset	Algorithms	Accuracy
20	Heart disease database	Random Forest Decision Tree Logistic Regression	88% 81% 85%

METHODOLOGICAL APPROACH

Historical Medical Data

The research adopts a methodology using independent variables (*e.g.*, age, gender, medical history, symptoms) along with the dependent variable (CHD class) which trains a model. This applied to accurately forecast CHD in test datasets. Using historical medical data from the South African Heart Disease: Knowledge Extraction Based on Evolutionary Learning (Fig. **1**) shows the methodology. The initial dataset undergoes preprocessing, eliminating noise and addressing missing values.

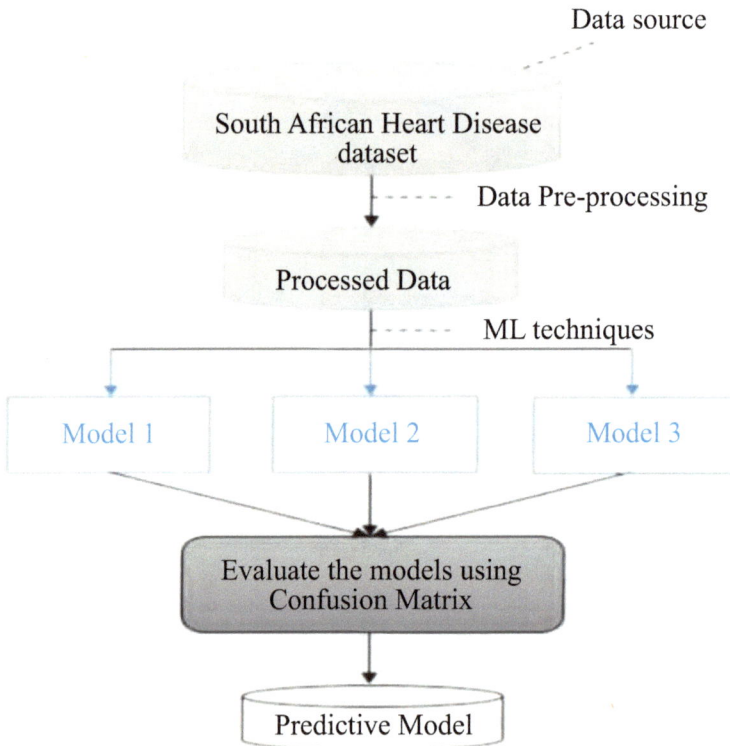

Fig. (1). Proposed methodology.

Using phonocardiogram (PCG) audio recordings, the proposed method combines two main components: classical ML and end-to-end DL to detect CHF. The traditional machine learning component includes segment-based machine learning, selection, and feature extraction. On the other hand, the end-to-end DL component analyses the raw PCG signal as well as its spectrogram representation using a deep Spectro-temporal ResNet architecture.

In the classic ML component, preprocessing involves a low-pass Butterworth filter and sliding window segmentation. OpenSMILE is employed for feature extraction, generating 1941 segment-based and 1941 recording-based features. Mutual information is used for efficient feature selection, and a Random Forest algorithm is trained as a segment-based classifier on the selected features.

The deep learning component employs a Spectro-temporal ResNet architecture, utilizing two branches for the time and frequency domains. The architecture is trained with binary cross-entropy loss and an Adam optimizer for 20 epochs. The network learns a Spectro-temporal encoding, providing valuable features for CHF detection.

The outputs from both ML and DL components are averaged for each recording. These aggregated features serve as input to a recording-based ML model, specifically a Random Forest classifier (RCF). This integration aims to capture informative aspects from both classic ML and end-to-end DL approaches, improving overall classification accuracy.

The first step in our process is segmenting heart sounds using Springer's technique to accurately distinguish different segments. Next, we extract twenty traits in total, each of which is linked to a different condition of the heart. We use random forest and logistic regression classifiers as comparison benchmarks to evaluate the performance of our suggested method. This baseline sets a benchmark for assessment and acts as a point of comparison to see how effective our new approach is.

During the training phase, three classifiers—segment-based, spectro-temporal ResNet, and recording-based—are trained independently. As a meta-learner, the recording-based classifier is trained using a stringent 10-fold cross-validation procedure. This guarantees that it can efficiently adjust to a variety of data circumstances. To validate the generalizability and robustness of our approach, we measure its performance on a separate test set that was not included in the training data, which yields a trustworthy indicator of its efficacy in a variety of unknown scenarios.

The proposed method demonstrates notable success, achieving an accuracy of 92.9% and outperforming baseline methods. It showcases robustness across diverse heart-related conditions and positions, making it a promising approach for comprehensive CHF detection. Classic ML and DL combinedly provide a synergistic advantage, offering the potential for personalized models and timely intervention in different CHF phases.

DECISION TREE ALGORITHM

The DT algorithm, a staple in supervised learning, tackles classification problems by structuring features as nodes and outcomes as leaves in a tree-like form. It simplifies decision-making processes and aids in interpretability (Fig. **2**).

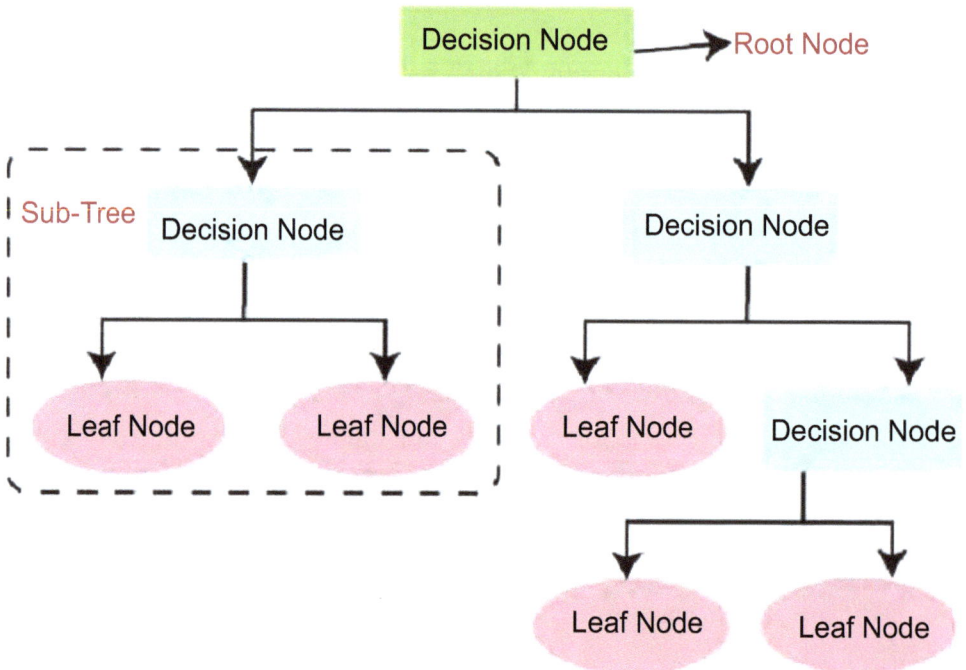

Fig. (2). DT Algorithm.

This is based on the CART algorithm, which optimizes decisions at each node for tasks involving classification and regression, thereby shaping the DT. CART strategically constructs the tree structure based on dataset features, providing a transparent and intuitive model for effective decision-making in various applications.

MACHINE LEARNING ALGORITHMS

The implementation of ML algorithms in this study centered on the WEKA tool, specifically version 3.8.2, chosen for its versatility and user-friendly graphical interface. The study conducted extensive testing, applying over twenty machine-learning algorithms within WEKA to datasets from the UCI Machine Learning Repository and Statlog Database dedicated to heart disease prediction. The selection process emphasized accuracy, resulting in the identification of the top five performers, all of whom achieved accuracy rates exceeding 80%, for further consideration.

- **Naïve Bayes:** Functioning as a statistical classifier, Naïve Bayes operates under the assumption of no attribute dependencies, determining class probabilities based on conditional independence. Its simplicity and minimal error rate contribute to its prominence.
- **Artificial Neural Networks (ANNs):** These biologically inspired models excel at capturing complex non-linear functions. Comprising layers for input, output, and hidden patterns, ANNs are particularly adept at recognizing intricate patterns within data.
- **Support Vector Machine (SVM):** Specializing in classifying both linear and non-linear data, SVM employs hyperplanes to effectively separate input variable spaces, maximizing the margin between different classes for improved accuracy.
- **Random Forest:** As a potent algorithm utilizing bagging (Bootstrap Aggregation), Random Forest constructs multiple decision tree models on diverse samples, enhancing prediction accuracy by aggregating their outputs.
- **Simple Logistic Regression:** Derived from statistical principles, LR is well-suited for binary classification tasks. It calculates coefficients for input variables and predicts output probabilities using a logistic function, providing interpretability and insights into the likelihood of outcomes.

These algorithms were essential to the study's assessment of the heart disease prediction models since they provided insightful information about the models' suitability, efficacy, and performance in relation to the study's goals. The emphasis on accuracy and the selection of the top-performing algorithms underscores the robustness of the chosen methodologies in the context of heart disease prediction.

The proposed methodology aims to utilize the ANN for heart disease prediction using 297 datasets out of 300 from the heart disease dataset. The first part involves identifying risk features by analyzing p-values for each feature. Subsequently, the dataset is divided for testing and training, and the trained dataset is input into the neural network, as illustrated in Figs. (**3** & **4**).

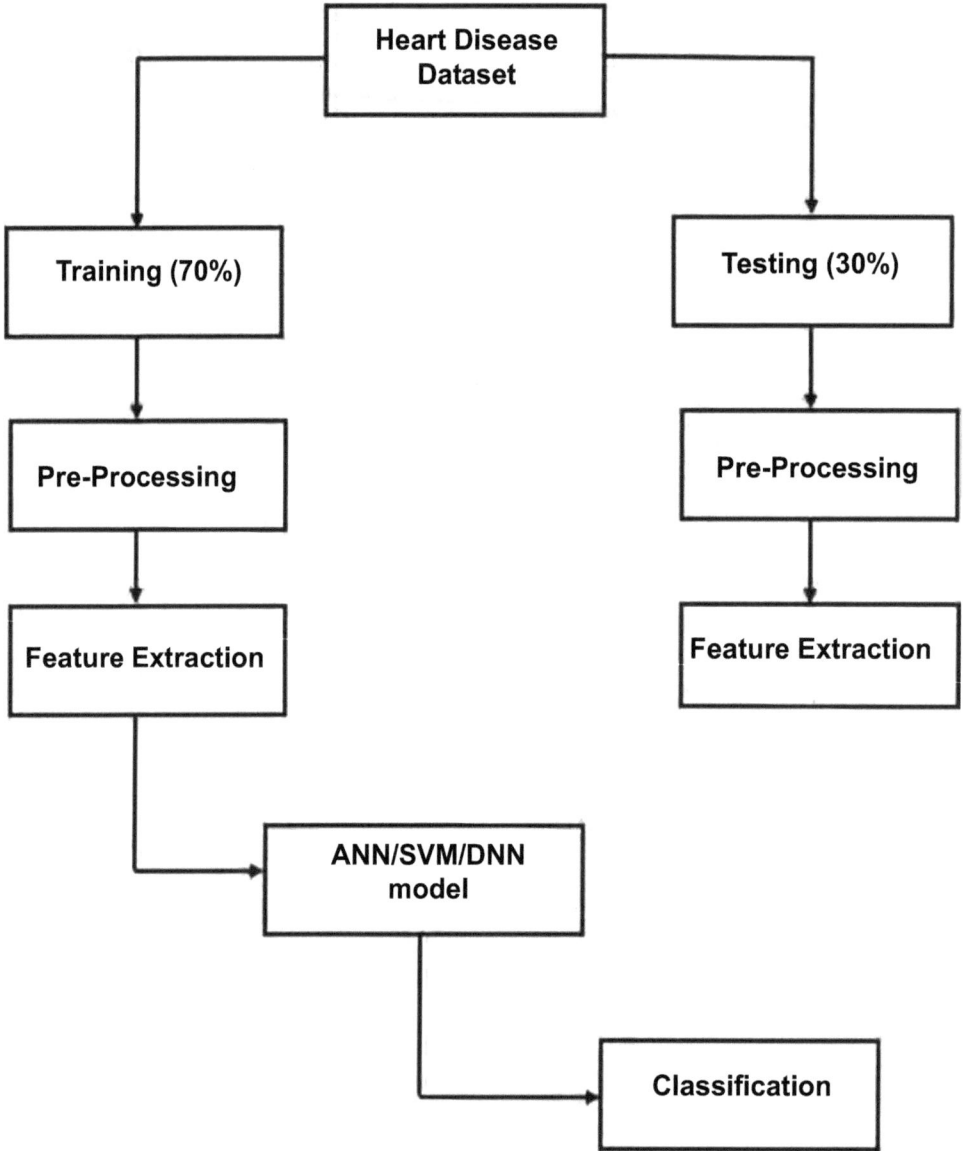

Fig. (3). Proposed architecture scheme.

Algorithm 1 Decision Tree-Based Partition

Require: Input: D dataset – features with a target class

   for $\forall$features **do**

     for Each sample **do**

       Execute the Decision Tree algorithm

     end for

     Identify the feature space $f_1, f_2, \ldots, f_x$ of dataset UCI. (9)

   end for

   Obtain the total number of leaf nodes $l_1, l_2, l_3, \ldots, l_n$ with its constraints (10)

   Split the dataset D into $d_1, d_2, d_3, \ldots, d_n$ based on the leaf nodes constraints. (11)

   Output: Partition datasets $d_1, d_2, d_3, \ldots, d_n$

DATA ANALYSIS

This study utilized the Java-based open-access data mining platform (WEKA) to select an efficient ML algorithm for heart disease prediction. The chosen algorithm's accuracy and performance were compared with those of other data mining algorithms. A smartphone app that predicted heart disease based on the cloud was then created using the top-performing algorithm. The application facilitates patient input of heart disease parameters and sensor data, storing the information on a cloud-based server. Additionally, a wireless patient monitoring system using Arduino and sensors collects real-time physiological data, triggering alerts and live video calls to doctors in critical situations. This comprehensive approach enhances remote patient monitoring and facilitates timely medical intervention.

RESULTS AND DISCUSSIONS

The use of machine learning (ML) algorithms, particularly CNN and GAN, is explored for cardiac disease prediction, with SVM emerging as a robust choice. The study focuses on coronary heart disease prediction using variables like age and gender. Experimental results in WEKA demonstrate the efficacy of ML techniques results are shown in Figs. (**4** and **5**). Another study combines classic ML and end-to-end DL for heart failure detection, achieving 92.9% accuracy. Both studies underscore the potential of AI and ML in cardiac diagnostics, emphasizing early detection and personalized approaches for effective management.

Fig. (4). The output of not having heart disease.

Fig. (5). The output of having heart disease.

Table **2** shows the results of the performance metrics of different ML models on a classification task. It calculates the precision, recall, F1-score, and support with the help of the accuracy of different models with their weighted metrics.

Table 2. Results obtained through models.

Sr. No.	Model	Precision	Recall	F1-score	Support
1	LR	0.89	0.875	0.88	138
2	RFC	0.89	0.875	0.88	138

(Table 2) cont.....

Sr. No.	Model	Precision	Recall	F1-score	Support
3	DT	0.81	0.81	0.81	138
4	Accuracy-LR	0.88	0.87	0.87	276
5	Accuracy-RFC	0.87	0.88	0.88	276
6	Accuracy-DT	0.82	0.81	0.81	276
7	Macro-avg-LR	0.87	0.86	0.86	276
8	Macro-avg-RFC	0.89	0.88	0.88	276
9	Macro-avg-DT	0.81	0.81	0.81	276
10	Weighted-avg -LR	0.87	0.87	0.87	276
11	Weighted-avg -RFC	0.89	0.88	0.88	276
12	Weighted-avg -DT	0.81	0.81	0.81	276

CONCLUSION

ML applications in heart disease diagnosis reveal the significance of real-time patient data and interpretable predictions, with a focus on deep learning, especially utilizing SMOTE for imbalanced data. The emergence of generative adversarial networks (GANs) is noted, with acknowledgment of computational costs. Another study on CHD prediction using SVM, DT, and NB highlights the potential for ML models in early detection, showcasing Naïve Bayes as the most accurate. A novel chronic heart failure (CHF) detection method blending ML with DL outperforms baselines and explores personalized models for distinguishing CHF phases. The application of seven ML algorithms on a dataset with 76 features, including SVM and Extreme Gradient Boosting, demonstrates the technology's potential in supporting early diagnosis, with Extreme Gradient Boosting being the most accurate. Emphasizing the critical importance of accurate heart disease prediction, a hybrid HRFLM approach combining random forest and linear method characteristics shows commendable accuracy. Lastly, a heart disease detection model with 94.96% accuracy, led by the Random Forest classifier, becomes a valuable tool for consistent diagnoses and efficient patient screening, with future research aiming to enhance predictive capabilities through diverse algorithms and parameters.

GAP ANALYSIS AND NEED OF RESEARCH

In order to diagnose cardiovascular disease, ML models frequently rely on intricate deep learning architectures, such as CNNs and DNNs, which are renowned for their excellent accuracy but difficult interpretability. For clinicians to trust and effectively use these models in medical settings, interpretability is essential. ML models that strike a balance between interpretability and accuracy should be the main focus of future research; this could be achieved by feature importance analysis or model-agnostic approaches. Furthermore, ensuring that machine learning models generalize effectively across a range of populations requires addressing the inherent bias in datasets, especially those that are skewed towards particular demographics. In order to improve model fairness and reliability across various patient demographics and geographic regions, more research is needed on techniques to mitigate bias during the data preprocessing and model training phases.

Despite the potential of machine learning models, integrating them into clinical workflows is still difficult. Broad acceptance by healthcare providers depends on seamless integration with current diagnostic systems and electronic health records (EHRs). To maximize clinical decision support systems' usefulness and efficacy in actual medical settings, future research should concentrate on creating user-friendly interfaces and integrating ML models into these systems. It is imperative to guarantee that machine learning models are reproducible and validated in a variety of datasets and healthcare settings. These models can be made more applicable and reliable through external validation with a variety of datasets, which will also support their strong performance across a range of patient cohorts and therapeutic scenarios. To improve patient outcomes and healthcare efficiency, more research should be done on IoT-based monitoring systems for real-time patient data collection and ML-driven interventions. These systems could greatly improve continuous monitoring and early detection of cardiovascular diseases. When implementing machine learning models in healthcare settings with limited resources, scalability, and cost-effectiveness are critical factors that should be taken into account. This calls for research into the best ways to maximize computational resources and assess the economic viability of these models in a variety of healthcare settings.

LIMITATIONS

Despite the promising results, certain limitations exist. The need for future research is highlighted to enhance model performance, specifically in terms of improving sensitivity and specificity rates. One suggested avenue for improvement involves exploring unsupervised learning techniques before making predictions. These limitations underscore the ongoing challenges and the potential for further refinement in utilizing ML for CHD prediction.

REFERENCES

[1]   L. Ali, A. Niamat, J. A. Khan, N. A. Golilarz, X. Xingzhong, and A. Noor, , "An optimized stacked support vector machines based expert system for the effective prediction of heart failure", *IEEE Access,* vol. 7, pp. 54007-54014, 2019.
        [http://dx.doi.org/10.1109/ACCESS.2019.2909969]

[2]   A. Rairikar, V. Kulkarni, V. Sabale, H. Kale, and A. Lamgunde, "Heart disease prediction using data mining techniques", *International Conference on Intelligent Computing and Control (I2C2),* p. 23-24, 2017

[3]   D. E. Salhi, A. A. K. Tari, and M. T. Kechadi, "Using machine learning for heart disease prediction," *Advances in Computing Systems and Applications*, pp. 70-81, 2021.

[4]   A. H. Gonsalves, F. Thabtah, G. Singh, and R. M. A. Mohammad, "Prediction of coronary heart disease using machine learning: An experimental analysis," *International Conference on Computer Information Science (ICCIS)*, pp. 5–7, 2019.
        [http://dx.doi.org/10.1145/3342999.3343015]

[5]   M. Gjoreski, "Machine learning and end-to-end deep learning for the detection of chronic heart failure from heart sounds", *IEEE Access,* vol. 8, pp. 16039-16049, 2020.
        [http://dx.doi.org/10.1109/ACCESS.2020.2968900]

[6]   J. S. Krishnan , and S. Geetha, "Prediction of heart disease using machine learning algorithms", *1st Int. J. Eng. Res. Technol. (ICIICT)* 2019.

[7]   P. K. Bhunia, A. Debnath, P. Mondal, M. D. E., K. Ganguly, and P. Rakshit, "Heart disease prediction using machine learning," *Int. J. Eng. Res. Technol. (IJERT)*, 2021.

[8]   V. Sharma, S. Yadav, and M. Gupta, "Heart disease prediction using machine learning techniques", *2nd International Conference on Advances in Computing, Communication Control and Networking (ICACCCN),* pp. 18-19, 2020.
        [http://dx.doi.org/10.1109/ICACCCN51052.2020.9362842]

[9]   A. Gavhane, G. Kokkula, I. Pandya, and K. Devadkar, "Prediction of heart disease using machine learning", *Proceedings of the 2nd International Conference on Electronics, Communication and Aerospace Technology (ICECA),* pp. 28-31, 2018.

[10]   A. Nikam, S. Mantri, S. Bhandari, and A. Mhaske, "Cardiovascular disease prediction using machine learning models", *Pune Section International Conference (PuneCon),* Vishwakarma Institute of Technology: Pune India, 2020.
        [http://dx.doi.org/10.1109/PuneCon50868.2020.9362367]

[11]   C S Dangare, and S S Apte, "Improved study of heart disease prediction systems using data mining classification techniques", *Int. J. Comput. Appl.,* vol. 47, no. 10, pp. 44-48, 2012.

[12]   A. Hazra, S.K. Mandal, A. Gupta, A. Mukherjee, and A. Mukherjee, "Heart disease diagnosis and prediction using machine learning and data mining techniques: A review", *Adv. Comput. Sci. Technol.,* vol. 10, no. 7, pp. 2137-2159, 2017.

[13]   K. Kant and K. Garg, "Review of heart disease prediction using data mining classifications," *IJSRD*, Vol. 2, no. 04, 2014, pp. 109-111.

[14]   M. Gandhi and S. N. Singh., , "Predictions in heart disease using techniques of data mining," *ICFTCAKM*, 2015, pp. 520-525.

[15]   R. Yaswanth and Y. M. Riyazuddin, "Heart disease prediction using machine learning techniques," *Int. J. Innov. Technol. Explor. Eng. (IJITEE)*, Vol 9, no. 5, 2020.

[16]   M. M. Ahsan and Z. Siddique, "Machine learning-based heart disease diagnosis: A systematic literature review, 2021.

[17]   S. Nashif, M. R. Raihan, M. R. Islam, and M. H. Imam, "Heart disease detection by using machine learning algorithms and a real-time cardiovascular health monitoring system", *World J. Eng. Technol.,* vol. 6, no. 4, pp. 57-64, 2018.
[http://dx.doi.org/10.4236/wjet.2018.64057]

[18]   M. Laad, K. Kotecha, K. Patil, and R. Pise, "Cardiac diagnosis with machine learning: A paradigm shift in cardiac care", *Appl. Artif. Intell.,* vol. 36, no. 1, pp. 1-14, 2022.

[19]   H. Meshref, "Cardiovascular disease diagnosis: a machine learning interpretation approach", *ACSA Int. J. Adv. Comput. Sci. Appl.,* vol. 10, no. 12, p. 258, 2019.
[http://dx.doi.org/10.14569/IJACSA.2019.0101236]

[20]   R. Pandey, and M. Choudhary, "Cardiovascular imaging using machine learning: A review", *Int. J. Recent Technol. Eng. (IJRTE),* vol. 11, no. 6, 2023..
[http://dx.doi.org/10.35940/ijrte.F7480.0311623]

[21]   M.A. Jabbar, B.L. Deekshatulu, and Priti Chandra, "Intelligent heart disease prediction system using random forest and evolutionary approach," *J. Netw. Innov. Comput.,* vol. 4, pp. 175-184, 2016.

[22]   V. V. Ramalingam, A. Dandapath, and M.K. Raja, "Heart disease prediction using machine learning techniques: a survey", *Int. J. Eng. Technol.,* vol. 7, no. 2.8, pp. 684-687, 2018.

[23]   S. Nikhar, and A.M. Karandikar, "Prediction of heart disease using machine learning algorithms", *Int. J. Adv. Eng. Manag. Sci.,* vol. 2, no. 6, p. 239484, 2016.

[24]   A. Nikam, S. Mantri, S. Bhandari, and A. Mhaske, "Cardiovascular disease prediction using machine learning models", *IEEE Pune Section International Conference (PuneCon),* Vishwakarma Institute of Technology: Pune India, 2020.
[http://dx.doi.org/10.1109/PuneCon50868.2020.9362367]

[25]   A. Rairikar, V. Kulkarni, V. Sabale, H. Kale, and A. Lamgunde, "Heart disease prediction using data mining techniques," *IEEE ICCC,* pp. 1-8, 2017.

CHAPTER 7

IoT Applications in Digital Health Care

Rohini Chavan[1,*] and **Shreyash Shabadi**[1]

[1] *E & TC Department, Vishwakarme Institute Of Technology, Pune, India*

Abstract: Over and past decade, significant advancements have been made in healthcare services, driven by continuous research and technological innovation. The Internet of Things (IoT), in particular, has demonstrated immense potential by connecting medical devices, sensors, and healthcare professionals to deliver high-quality care even in remote and isolated areas. This technology has enhanced patient safety, enabled at-home care, reduced healthcare costs, improved service accessibility, and increased operational efficiency within the healthcare industry. IoT has transformed the traditional hospital-centered healthcare system into a patient-centered model. Various clinical analyses, such as monitoring blood pressure, blood glucose levels, and oxygen saturation (pO_2), can now be performed at home without requiring assistance from healthcare professionals. Additionally, clinical data collected remotely can be transmitted to healthcare centers using advanced telecommunication technologies. The integration of these communication services with IoT has significantly improved access to healthcare facilities.

This study provides a comprehensive overview of the latest IoT-based healthcare applications, focusing on enabling technologies, healthcare services, and solutions to address various healthcare challenges. Furthermore, it serves as a valuable resource for future researchers interested in contributing to the advancement of IoT technologies in patient care, offering insights into the diverse applications and benefits of IoT in the healthcare sector.

Keywords: Clinical analysis, Centric healthcare, Healthcare services, Healthcare professionals, Internet of Things (IoT), Patient care.

INTRODUCTION

The fusion of healthcare and cutting-edge technology has given rise to a transformative revolution in the way we perceive and practice medicine. In this era of unprecedented innovation, the Internet of Things (IoT) has emerged as a pivotal force in reshaping healthcare as we know it. The amalgamation of IoT and

---

* **Corresponding author Rohini Chavan:** E & TC Department, Vishwakarme Institute Of Technology, Pune, India;
E-mail: rohini.chavan@viit.ac.in

Parikshit N. Mahalle, Gitanjali R. Shinde, Namrata N. Wasatkar & Prashant R. Anerao (Eds.)

healthcare often referred to as "IoT in Healthcare" or "Healthcare IoT," is not merely a technological leap; it represents a profound shift towards more personalized, efficient, and accessible healthcare solutions.

IoT, at its core, involves the interconnection of everyday objects and devices to the internet, enabling them to collect, exchange, and analyze data autonomously. In the realm of healthcare, this concept transcends conventional boundaries, creating a networked ecosystem where medical devices, wearables, sensors, and patient data converge to empower both healthcare providers and patients alike. The architecture of IoT-based health care system is given in Fig. (**1**).

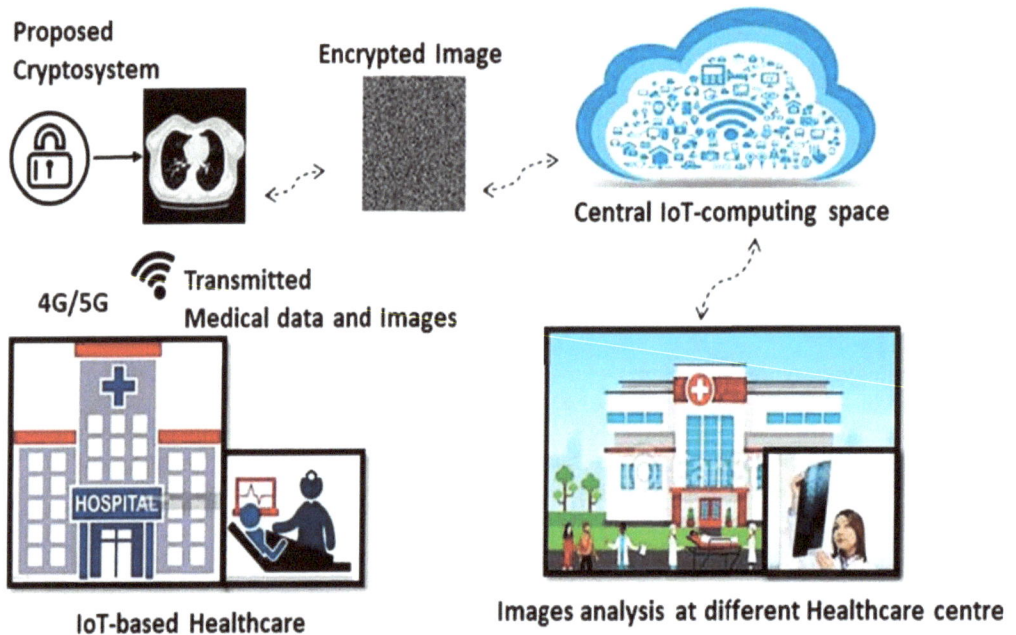

Fig. (1). The architecture of IoT-based health care system [1].

This article embarks on a journey through the intricate landscape of IoT applications in healthcare, exploring the multifaceted ways in which this technology is reshaping the healthcare landscape. We will delve into the innovative solutions that IoT offers, from remote patient monitoring and smart medical devices to predictive analytics and telemedicine. Moreover, we will examine the broader implications of these advancements, including improved patient outcomes, cost reduction, and the democratization of healthcare.

As we navigate this captivating intersection of technology and healthcare, it becomes apparent that IoT applications have the potential to not only enhance the quality of care but also usher in a new era of healthcare delivery, one that is more patient-centric, data-driven, and interconnected than ever before. In doing so, IoT in healthcare is not merely a technological advancement but a profound catalyst for the betterment of human health, promising a future where healthcare is not just reactive but predictive and preventive, where patient well-being is constantly monitored, and where healthcare services are tailored to individual needs. Welcome to the world of IoT applications in healthcare, where the future of medicine is taking shape before our eyes.

Over the past decade, significant research and technological advancements have been made in healthcare services. The Internet of Things (IoT) has demonstrated promising applications by connecting various medical devices, sensors, and healthcare professionals to deliver high-quality medical care in remote and inaccessible areas. IoT has contributed to enhancing patient safety, reducing healthcare costs, improving access to healthcare services, and increasing operational efficiency within the healthcare industry.

This study provides a precise overview of the potential applications of IoT-based technologies in healthcare. It highlights the advancements in Healthcare IoT (HIoT) from the perspectives of enabling technologies, healthcare services, and applications aimed at addressing various healthcare challenges.

Technology has transformed healthcare from a hospital-centric system to a patient-centric model [2, 3]. For instance, various clinical tests, such as monitoring blood pressure, blood glucose levels, and pO_2 levels, can now be conducted at home without requiring assistance from healthcare professionals. Clinical data gathered in remote areas can be transmitted to healthcare centers using advanced telecommunication services.

The integration of these services with rapidly evolving technologies, including machine learning, big data analytics, IoT, wireless sensing, mobile computing, and cloud computing, has significantly improved access to healthcare facilities. IoT, in particular, has broadened human interaction with the external environment by utilizing advanced protocols and algorithms. It connects numerous devices, wireless sensors, home appliances, and electronic systems to the Internet [1].

The growing popularity of IoT is attributed to its benefits, such as increased accuracy, reduced costs, and enhanced ability to predict future events. The rapid adoption of IoT has been fueled by advancements in software and applications, the widespread availability of wireless technologies, and the expanding digital economy, leading to an accelerated IoT revolution [4].

Healthcare applications utilize various sensors to collect physiological data from patients, such as temperature, heart rate, electrocardiograph (ECG), and electroencephalograph (EEG). Additionally, environmental factors like temperature, humidity, date, and time can also be recorded. These data types are crucial for deriving accurate and meaningful insights into patients' health conditions. Data storage and accessibility play a significant role in IoT systems, as they must handle vast amounts of information gathered from diverse sources, including sensors, mobile phones, emails, software, and applications [5]. The collected data from these sensing devices is shared with doctors and authorized personnel for further diagnosis. It is transmitted to healthcare providers *via* the cloud, enabling rapid diagnosis and timely medical intervention if necessary. Effective collaboration between doctors, patients, and communication modules is essential to ensure the secure and efficient transmission of information [6].

Most IoT systems employ a user interface that serves as a dashboard for medical caregivers, facilitating user control, data visualization, and interpretation. Extensive research in the literature highlights advancements in IoT systems for healthcare monitoring, control, security, and privacy [7].

IOT APPLICATIONS IN HEALTH CARE

The Internet of Things (IoT) has significant applications in healthcare. IoT applications are used for enhancing patient care, improving efficiency, and reducing costs. Here are some key IoT applications in healthcare:

Activity Trackers During Cancer Treatment

Determining the appropriate treatment for a cancer patient requires considering more than just their weight and age. Factors such as lifestyle and fitness levels significantly influence the development of an effective treatment plan. Activity trackers play a crucial role in monitoring a patient's movements, fatigue levels, appetite, and more. The data collected by these trackers, both before and after treatment begins, provides healthcare professionals with valuable insights to make necessary adjustments to the recommended treatment plan [8].

Heart Monitors with Reporting

Patients can use wearable devices that monitor their heart rates and detect conditions such as high blood pressure. These devices enable healthcare providers to access heart monitor data during checkups and examinations as needed. Additionally, wearable devices can alert healthcare professionals to critical conditions such as arrhythmias, palpitations, strokes, or heart attacks. This timely notification allows ambulances to be dispatched promptly, potentially making the difference between life and death [9].

Medical Alert Systems

Individuals can wear devices resembling jewelry, such as medical alert bracelets, designed to notify family members or friends in case of an emergency. For example, if someone wearing a medical alert bracelet falls out of bed during the night, their designated emergency contacts are instantly alerted *via* their smartphones, ensuring timely assistance [10].

Ingestible Sensors

Patients can now use ingestible devices equipped with sensors that resemble pills. Once swallowed, these sensors transmit data to a mobile app, helping patients adhere to the correct medication dosages. This technology addresses the common issue of medications not being taken as prescribed due to forgetfulness or human error. By providing reminders and tracking, the ingestible sensor ensures patients take the right medications, at the right time, in the correct dosages [11].

Trackable Inhalers

IoT-enabled inhalers provide patients with insights into potential triggers for asthma attacks by transmitting data to their smartphones or tablets. This information can also be shared with their physicians for better management. Additionally, these connected inhalers send reminders to patients about when to take their medications, helping to ensure proper usage [12].

Remote Patient Monitoring (RPM)

Wearable Devices: IoT-enabled wearable devices can continuously monitor vital signs such as heart rate, blood pressure, and glucose levels. This real-time data allows healthcare providers to track patients remotely and intervene if necessary.

Implantable Devices: Implantable IoT devices can monitor conditions internally, such as pacemakers that transmit data to healthcare providers for remote monitoring [13].

Smart Health Records

IoT helps in the creation and maintenance of electronic health records (EHRs) by collecting and updating patient data in real time. This improves the accuracy and accessibility of patient information for healthcare professionals [14].

Medication Adherence

IoT devices can be used to monitor and improve medication adherence. Smart pill dispensers can send reminders to patients to take their medication and notify healthcare providers if doses are missed [15].

Asset and Inventory Management

IoT facilitates the tracking of medical equipment, medications, and other assets in healthcare facilities. This helps in reducing waste, improving efficiency, and ensuring that necessary supplies are always available [16].

Telemedicine and Telehealth

IoT is essential in facilitating remote consultations and telehealth services. Connected devices enable healthcare providers to monitor patients, conduct virtual appointments, and deliver timely interventions, all without requiring physical presence [17].

Fall Detection and Prevention

IoT sensors can be integrated into healthcare facilities and homes to detect falls or other emergencies, especially in the case of elderly patients. This allows for quick response and reduces the risk of injuries [18].

Environmental Monitoring

IoT devices can monitor environmental conditions in healthcare facilities, ensuring optimal conditions for patient comfort and safety. This includes monitoring temperature, humidity, and air quality [19].

Predictive Analytics

By analyzing data from IoT devices, healthcare providers can use predictive analytics to detect potential health problems before they escalate. This proactive strategy helps improve disease management and prevention [19].

Chronic Disease Management

For patients with chronic conditions, IoT devices can assist in continuous monitoring, allowing healthcare providers to adjust treatment plans based on real-time data, leading to more personalized and effective care [19].

Clinical Trials and Research

IoT devices are used in clinical trials for remote data collection, monitoring patient response to treatments, and ensuring protocol compliance. This can streamline the research process and improve the accuracy of results [20].

Big Data Could Cure Cancer

Big data can cure cancer using big data, a system called "Cancer Moonshot" has been developed that helps find a cure for cancer. The program aims to treat cancer in a shorter time than traditional methods. Many researchers in the medical field are using large amounts of data to obtain better results and identify beneficial effects. The results of this system will give good results to the treatment plan. Additionally, tumor samples stored in the biobank are linked to medical records. Sequentially calculating information about the interaction between cancer cells and mutations means treatment plans can be improved and lead to better outcomes. Perhaps this could lead to surprising results, such as desipramine, which has been shown to treat lung cancer. These organizations have medical records.

The genetic sequencing of 13 of 19 cancer cells taken from experimental patients was performed and added to the World Cancer database. However, there are some limitations in the use of big data analytics to analyze cancer treatment, mainly due to the problem of inconsistent data, as some data records are not associated with other records. Predictive Analytics in HealthcarePredictive Analytics has been used in business applications for recent years and will be further researched in the future. A recent study showed that a US research project collected data on 30 million people and monitored birth quality [12]. Doctors and therapists make decisions based on available information, providing better and more effective treatment. It might be more useful. There are also new tools that can predict heart disease, high blood pressure, and diabetes. Patients suffering from these conditions may be advised to attend regular physical examinations, receive regular medical care, and use a diet or weight management plan. The proposed method has been compared in terms of accuracy with many existing systems for cardiovascular disease prediction, such as GV + SVM, SVM, MLP, ANN, LR, KNN, NB, DT, SVM + MLP. The analysis results of the IoT method show a reality that it has the potential to be used in medicine. Protein means treatments can be improved for better results. Perhaps this could lead to surprising results, such as desipramine, which has been shown to treat lung cancer. These organizations have medical records. Cancer samples taken from experimental patients were analyzed and added to the world cancer database. However, there are some limitations in the use of big data analytics to analyze cancer treatment, mainly due to the problem of data inconsistency, as some data cannot be correlated with other data. Predictive Analytics in Healthcare has been used in business applications in recent years and will certainly be explored in the future. A recent study showed that a US research project collected information on 30 million people and monitored birth quality. Doctors and therapists provide better, more effective treatment by making decisions based on available information. It might be more useful. There are also new tools that can predict heart disease, high blood pressure, and diabetes. Patients suffering from these conditions may be advised to attend regular physical exams, receive regular medical care, and use a diet or weight management plan. The proposed method has been compared in terms of accuracy with many existing multiple prediction systems for cardiovascular disease prediction [18, 19] such as GV + SVM, SVM, MLP, ANN, LR, KNN, NB, DT, and SVM + MLP as shown in Fig. (**2**). The results of the analysis of the IoT method showed high accuracy, indicating its potential use in the medical field.

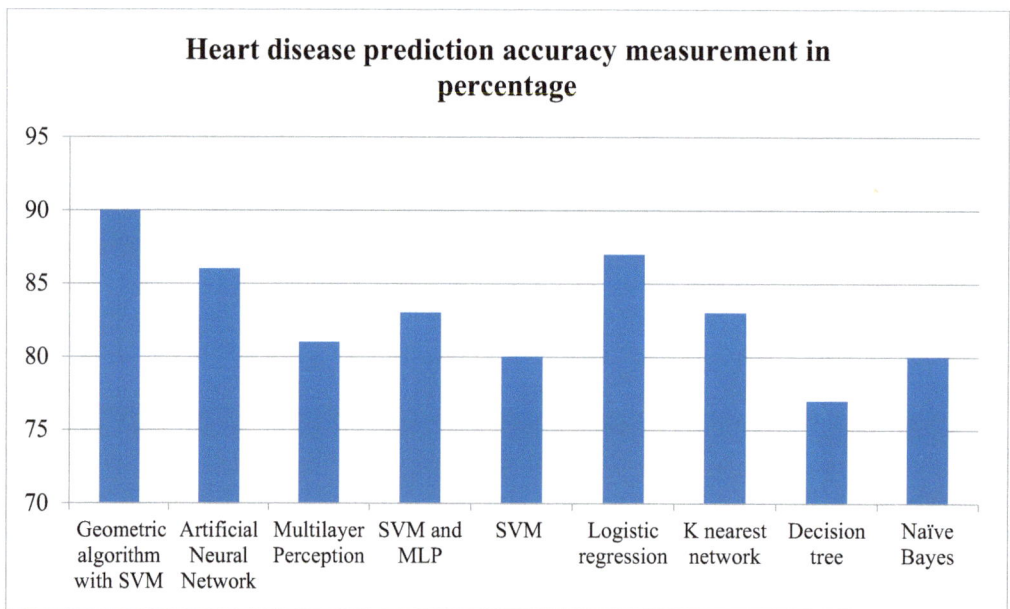

Fig. (2). Heart disease prediction accuracy across multiple prediction systems [6].

Big Data and Medical Imaging

Big data has made a great contribution to diagnosis; for example, many radiologists can obtain information about medical imaging diseases, identify them, and keep them accessible for years. However, this method is more expensive and takes longer. These numbers play an important role in diagnosis and thus help doctors treat diseases. This idea saves doctors from missing many medical images, which are very difficult because they contain detailed information. Therefore, the Carestream algorithm has a positive impact on the healthcare industry. The research and H-IoT solutions and challenges in the medical industry are given in Table **1**.

Table 1. Survey on the prevailing challenges in the healthcare industries associated with the possible IoT solutions.

Sr. No.	Challenges	IoT Solutions	Author
1.	Lack of hospitals that are smart	• IoT (Internet of Things) uses private networks and automation to deliver smart hospitals. • Providing all information digitally can reduce patient waiting times.	[12, 14]
2.	Storage of patient data (COVID-19)	• IoT devices effectively transmit, store, and analyze patient data to provide better healthcare in the future. • This will help identify and analyze returning patients. • Raise awareness about the causes of COVID-19 increase.	[20, 23]
3.	Treating COVID-19 patients	• Various medical equipment such as pumps, scales, and antibiotics are used in the care of COVID-19 patients. • IoT devices have instant services and can be used to treat COVID-19 patients.	[17]
4.	Lack of accurate decision making	• IoT devices make it easier to make good decisions because they can continuously predict and monitor patient data to make decision.	[24]
5.	Improper patient monitoring	• IoT devices constantly monitor cleanliness and pollution in support areas and hospitals. • The daily life of the patient is monitored for health improvement.	[27]

CHALLENGES IN IOT HEALTHCARE

Specific examples of the use of IoT in healthcare currently include symptom monitoring, medical education, wellness and collaborative disease management, and social care. Analyzing software applications can improve the interpretation of data and reduce the time required to reprocess generated data. The idea of learning big data will lead to the electronic evolution of medical disciplines, business processes, and timely decisions. As the world population increases, it is important to improve the understanding and interpretation of health and wellness information, reduce chronic diseases and vegetable consumption, eat well, improve mental health, and improve health and lifestyle. While it is not possible to list all IoT medical applications, we will provide an overview of the main ones. By taking a look at the scientific data and some studies, it is clear that IoT would soon play an important role in cancer treatment, patient self-assessment process, medication distribution, and care compliance, exaggerated severe pain management. Cancer Care (CC) wearables are clinically tested. The clinical trial was presented at the American Society of Clinical Oncology Annual Meeting in 2018. The study focused on head and neck cancer patients monitored with Bluetooth-enabled scales and blood pressure cuffs, as well as a symptom-tracking app that did not regularly send updates to patients. Approximately 400 patients participated in the study, and patients using this IoT system had fewer symptoms compared to the control group who received a physical examination every week [15, 16, 20].

In IoT diabetes medication, Insulin injection is a disease model used to measure self-care and health care based on a variety of domains, such as measuring blood sugar and measuring blood pressure. IoT-based continuous glucose monitoring can be done on many existing devices. While patients with type 1 diabetes (T1D) often require ongoing care and emergency intervention, growing evidence suggests that greater and regular monitoring over time may prevent complications in T2D patients [21, 22]. Smart insulin pens are useful tools to measure treatment compliance in patients with diabetes (DM). Although existing equipment is mostly used for insulin injections, similar equipment can also be used for medication containers. Nowadays, such wearable devices are connected to smartphones and doctors constantly monitor them. Integrating these models into an IoT environment allows doctors to be more quickly informed about patients who neglect treatment and compliance [23, 24]. Closed-loop (automated) insulin delivery systems have long been thought to be very effective in T1D care. Possible policies and mismanagement are affecting the description of products in healthcare. We have seen some suggestions from the network of doctors and patients considering that IoT could be useful to solve these problems. Although several steps are required, a mechanical and IoT-safe closed loop is essential for 1D patients at risk of diabetic ketoacidosis [25]. Diabetes and asthma are chronic diseases with exaggerated patterns and provide fertile ground for IoT-based treatment. This is a heavy burden for hundreds of millions of people around the world. Most patients are young and active and looking for stability in life. IoT wearable devices that measure satiety or warn of the presence of allergens are important for early detection and management of potential outbreaks. In the same framework, IoT-based inhalers can provide reliable information to the patient's doctor about compliance and the patient's ability to accurately control the device [14]. Asthma is of course a chronic condition, like a mental illness. In addition to the monitoring options mentioned above, IoT can improve patient support services. Combined with artificial intelligence, IoT can support chatbots for many purposes, from suicide detection to cognitive rehabilitation for patients with dementia or mild mental retardation [11]. In this section, we present IoT applications in healthcare, such as cancer treatment, patient self-assessment programs, medication delivery and monitoring, and asthma management. This model demonstrates the potential of IoT to improve healthcare, patient management, and research if further approval is obtained.

These applications demonstrate the potential of IoT to transform healthcare by providing more personalized, efficient, and accessible services while improving patient outcomes and reducing healthcare costs.

IoT promises to revolutionize healthcare. However, there is no change without controversy. In fact, it is the conflict between methods and ethics that leads to a unified approach. The issues of IoT in healthcare can be divided into process, cost-related, and ethical [23].

Technical Challenges

These occur because IoT is not yet a part of daily life. In many countries, fifth-generation wireless technology (5G) and the next Internet of Things services are not yet available. Not only patients but also most doctors and researchers have little understanding of what IoT has and what it can do in daily life outside of the pain of treatment. This shows that 5G can be considered the first challenge for the use of IoT in healthcare [16]. The use of 5G requires the installation of a large number of antennas. This is both expensive, time-consuming, and harmful to health. Although there is not enough evidence to support this claim, more research is needed to determine the safety of using 5G technology in general. Regarding the global impact of 5G, convincing policymakers and the public to change (with its programmes) will take as much or more time than security research has envisioned. The expected benefits of using IoT in healthcare will be the main argument in favor of 5G payments [11, 18].

The next technical challenge is the integration of information. More space means more devices. There are many wearable devices and data collection devices in healthcare that cannot easily be converted into a single data collection system due to reasons and money. Even now companies have not agreed on communication protocols and standards [16]. Patients with the same disease may use different tools to cope with the disease itself or one's significant symptoms. For example, people with diabetes may use different insulin pumps as well as different blood sugar and vital signs monitors. This suggests that at least three types of data were taken together from the same patient. Although this document will eventually be completed, it will take more time and this will become a problem in emergency management [18].

Financial Difficulties

From a financial perspective, the Internet of Things is leading the way in telemedicine. According to the International Data Corporation(IDC), the current budget for telehealth services in Europe is 10.41 billion euros, reaching over 12.4 billion euros. This type of funding can be approved for the development of IoT healthcare services. However, the complexity of IoT implementation has deterred many potential entrepreneurs. Sources of financial uncertainty include the involvement of third-party service providers to ensure the quality of the IoT and continuous processes. Observations clearly show that neither the government nor private providers are willing to develop IoT healthcare services without the evidence and knowledge of other countries/countries [25]. To be clear, the size of the IoT healthcare market was estimated at $60 billion in 2014. It is expected to reach a value of US$ 136 billion by 2021. The compound annual growth rate (CAGR) of IoT, particularly in healthcare, is expected to reach or exceed 12.5% during the forecast period. Whether this change can be sustained depends on the ability of external and internal medicine providers to achieve and maintain adequate understanding and cooperation. If implemented, IoT has the potential to reduce the cost of healthcare. Health-related costs are generally divided into direct costs and indirect costs. The former takes into account costs to the provider, while the latter includes costs to the recipient, including absenteeism, unpaid medical bills, and involvement of family members or other caregivers in their treatment. Given that IoT healthcare has not yet been incorporated into a major healthcare system, there is little evidence of cost-effectiveness [7]. Economists have highlighted the potential benefits of IoT in healthcare on a sustainable basis. Asset management, inventory management, tight control, good packaging and supply chain management have been recognized as prerequisites for the financial success of the Internet of Things in healthcare. However, to date, such concepts are business-centric rather than medical-focused. Their transfer to treatment requires good cooperation and deep understanding between economists, healthcare providers, and doctors [10, 11].

Ethical Challenges

Discussions about the Internet of Things in healthcare stem from data management and maintenance paradigms. When it comes to managing health-related data, the key concepts are data privacy, data sharing and freedom, data ownership and consent, and unknown value issues. In nursing, the isolation and dehumanization of doctor-patient communication, the decontextualization of health and well-being, and the risk of unprofessionalism in care lead to all kinds of panic [22]. Ethical implications will influence legal policy regarding the Internet of Things in healthcare. All laws must comply with international and regional standards, such as the Universal Declaration of Human Rights and the European Union's General Data Protection Regulation (GDPR). In this context, policy development is also expected to be a potential source of ethics for the use of IoT in healthcare [13]. Although GDPR only applies to the European Union, it could impact IoT in health research. Analyzing some real-world situations in the Internet of Things in the context of healthcare is important for understanding ethics. Sensors that track people at work and home will become a part of daily life and may be forgotten by their users. However, disappearing into the background, the same sensors will monitor the user at every moment of his own life, including killing in battle or changing the pulse of happiness. Voice detection and analysis sensors will also "hear" users' private conversations. Even if users agree to do so for their own health, such surveillance can compromise the privacy of their families, friends, and colleagues [7, 23]. Researchers have found a promising memory option for IoT-connected sensors. We recognize that drawing the red line between what is private and confidential and what is medically important is a difficult task. Daily family discussions or quiet conversations about users' contradictory behavior can hide latent signs of hypertension or cardiac arrhythmia. There is a trained physician or data scientist who sees the importance of self-selection being a privacy violation if the sensor does not work [9]. Also, data collected by the individual and selected by sensors is analysed to decide which risk will lead to ethical behavior [24].

Cyber security is a part of IoT healthcare that should not be ignored. Data storage and processing require a cloud-based service. Even though all ethical considerations have been addressed regarding healthcare and external services providing access to this information, hacking is still very dangerous. Insurance companies and human resources agencies are undermining the ability to recover effectively. Employees will have biometric data and medical history breached. In the same context, any other person may make financial or other demands to avoid semantic sensitivity [25].

CRITICAL ANALYSIS

The chart shown in Fig. (**3**) gives an overview of data defined according to different encryption platforms.

Many research papers focus on the advantages of using blockchain technology in healthcare, but they fail to adequately address issues such as capacity building, interoperability, and security management-related relationships.

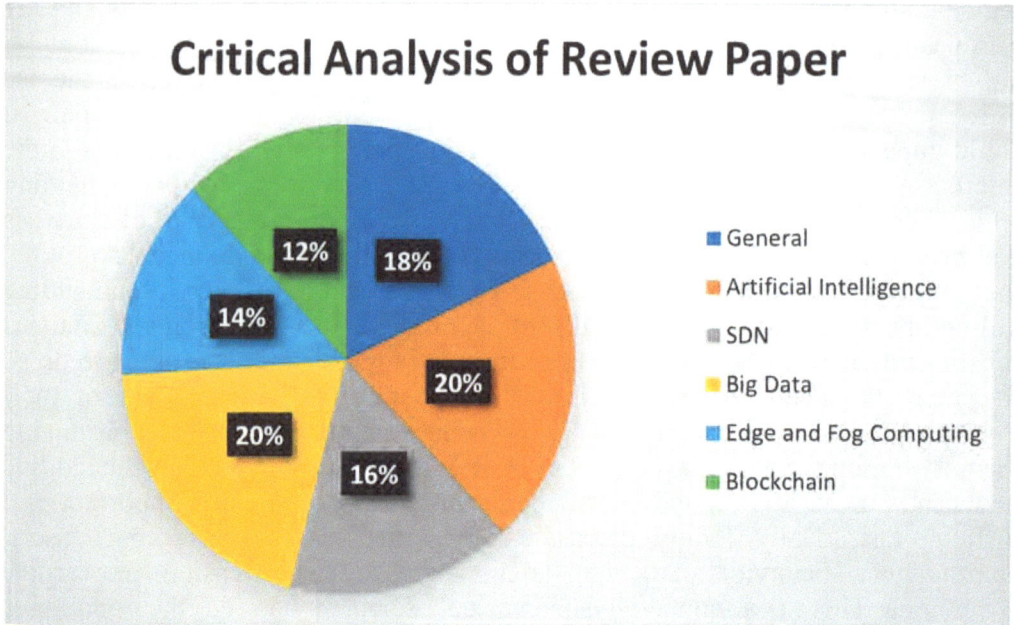

Fig. (3). Visualization of the different aspects of IoT [7].

Although many researchers have explored the potential of edge computing in healthcare, few have conducted a comprehensive assessment of its limitations. While edge computing can speed up data processing, it requires computing power and critical infrastructure that small healthcare organizations may struggle to obtain. Some researchers are interested in the potential uses of AI in medicine but have not yet conducted an in-depth analysis of the ethical limitations and implications of AI.

DISCUSSION

IoT technology collects and processes large amounts of data, helping to make timely and actionable decisions. In addition, smart tools have been developed to check how much the data has changed. The widespread use of the Internet of Things is one of the main benefits of technological reform. The Internet of Things will be one of humanity's greatest innovations. The popularity of the Internet of Things is a response to rapid growth over the last few hundred years. As we all know, important decisions need to take into account a lot of information about various points and things that affect the decision-making process. In fact, decisions need to be made almost instantly. People cannot make quick and correct decisions in such situations due to some physical and mental limitations. Therefore, IoT and AI technology are interrelated. The use of this technology can benefit the healthcare industry. Today's healthcare industry is an area where the Internet of Things is becoming popular. The Internet of Things directly affects people's lives and shows the importance of the Internet of Things. Medicine is a field of activity in daily life. The healthcare system is one of the most important areas where digital information is used a lot. Using interactive information and new technologies, it is not possible to process large data streams and turn them into final products and their products. According to a research study, most healthcare leaders believe that the Internet of Things will revolutionize medicine in the coming years. This will typically include three areas: remote health care for patients, chronic disease prevention, and data collection. Healthcare is the fastest-growing segment of the Internet of Things. According to scientists, the number of connected medical devices will increase 10-fold in the next 10 years. Many statistics estimate that the number of individual wearable devices has increased. It will increase to 92.1 million units in the next two years. In 2016, there were only 2.4 million people. At the same time, many devices and smart machines are designed not to replace doctors and nurses, but to support and enhance their work. Doctors can use online technology to help people remotely. This is especially important at a time when the epidemic is worsening. Therefore, IoT can find a way to treat every patient's disease. In addition, IoT can also help improve healthcare because automation of data collection in hospitals makes healthcare staff more efficient and provides more information. IoT can also improve prediction and disease prevention outcomes. Thanks to the relevance and integration of artificial intelligence and the Internet of Things, the number of medical errors will decrease, which will help save more patients. The emergence and widespread use of artificial intelligence in many areas of human activities has currently given rise to many debates. There are some problems and risks in the development of IoT technology. The main issue is a lack of understanding of the value of IoT. More research is needed to identify strategic measures to overcome these issues because they are the source of many threats when using IoT.

CONCLUSION

In conclusion, the integration of Internet of Things (IoT) technologies in healthcare has ushered in a new era of patient care, management, and overall system efficiency. The myriad applications of IoT in healthcare, from remote patient monitoring to smart health records and predictive analytics, offer tangible benefits for both healthcare providers and patients.

IoT devices facilitate real-time data collection, enabling healthcare professionals to monitor patients remotely and intervene promptly when needed. This not only enhances patient outcomes but also contributes to the shift towards proactive and personalized healthcare. Wearable devices, implantable sensors, and smart health records ensure that healthcare professionals have access to accurate and up-to-date patient information, leading to more informed decision-making.

The implementation of IoT in healthcare is not limited to patient care alone. Asset and inventory management, environmental monitoring, and predictive analytics contribute to the overall efficiency of healthcare facilities. Telemedicine, supported by IoT, extends the reach of healthcare services, particularly in remote or underserved areas.

Furthermore, the ability of IoT devices to collect and analyze vast amounts of data opens new avenues for research and clinical trials. This data-driven approach not only accelerates the pace of medical research but also enhances the precision and reliability of study outcomes.

While the benefits of IoT in healthcare are substantial, it is crucial to address challenges such as data security, privacy concerns, and interoperability standards. As the healthcare industry continues to embrace and refine IoT applications, collaborative efforts among stakeholders will be essential to ensure the seamless integration and secure management of these technologies.

In essence, the IoT's transformative impact on healthcare holds the promise of a more patient-centric, efficient, and responsive healthcare ecosystem. As technology continues to advance, the ongoing synergy between healthcare and IoT will likely drive further innovations, ultimately shaping a healthier and more connected future.

REFERENCES

[1]   A.G. Sreedevi, T. Nitya Harshitha, V. Sugumaran, and P. Shankar, "Application of cognitive computing in healthcare, cybersecurity, big data and IoT: A literature review", *Inf. Process. Manage.,* vol. 59, no. 2, p. 102888, 2022.
      [http://dx.doi.org/10.1016/j.ipm.2022.102888]

[2]   M.N. Bhuiyan, M.M. Rahman, M.M. Billah, and D. Saha, "Internet of Things (IoT): A review of its enabling technologies in healthcare applications standards protocols security and market opportunities", *IEEE Internet Things J.,* vol. 8, no. 13, pp. 10474-10498, 2021.
      [http://dx.doi.org/10.1109/JIOT.2021.3062630]

[3]   H. Raj, M. Kumar, P. Kumar, A. Singh, and O.P. Verma, "Issues and challenges related to privacy and security in healthcare using IoT fog and cloud computing", *Advanced Healthcare Systems: Empowering Physicians with IoT-Enabled Technologies,* Physicians IoT-Enabled Technol, pp. 21-32, 2022.
      [http://dx.doi.org/10.1002/9781119769293.ch2]

[4]   N. Jabeur, A. Yasar, Y. Mohamad, and M. Melchiori, "Guest editorial: Special issue on data science approaches and applications", *Comput. Inf.,* vol. 41, no. 1, pp. 1-11, 2022.
      [http://dx.doi.org/10.31577/cai_2022_1_1]

[5]   Gonçalo Marques, Akash Kumar Bhoi, Victor Hugo C. De Albuquerque, Hareesha K.S.," *IoT in Healthcare and Ambient Assisted Living, Studies in computational Intelligence", Studies in computational Intelligence*, vol. 933, Springer, 2021.

[6]   M. Kumar, A. Kumar, S. Verma, P. Bhattacharya, D. Ghimire, S. Kim, and A.S.M.S. Hosen, "Healthcare internet of things (H-IoT): Current trends, future prospects, applications, challenges, and security issues", *Electronics (Basel)*, vol. 12, no. 9, p. 2050, 2023.
[http://dx.doi.org/10.3390/electronics12092050]

[7]   N. Surantha, P. Atmaja, David, and M. Wicaksono, "A review of wearable internet-of-things device for healthcare", *Procedia Comput. Sci.*, vol. 179, pp. 936-943, 2021.
[http://dx.doi.org/10.1016/j.procs.2021.01.083]

[8]   S.J. Rigo, and S.D.L.M. Correia, "Machine learning and IoT applied to cardiovascular diseases identification through heart sounds: A literature review", *Proc. ICITS*, vol. 414, p. 356, 2022.

[9]   A. Shamsabadi, Z. Pashaei, A. Karimi, P. Mirzapour, K. Qaderi, M. Marhamati, A. Barzegary, A. Fakhfouri, E. Mehraeen, S. SeyedAlinaghi, and O. Dadras, "Retracted: Internet of things in the management of chronic diseases during the COVID-19 pandemic: A systematic review", *Health Sci. Rep.*, vol. 5, no. 2, p. e557, 2022.
[http://dx.doi.org/10.1002/hsr2.557] [PMID: 35308419]

[10]   G. Gopal, C. Suter-Crazzolara, L. Toldo, and W. Eberhardt, "Digital transformation in healthcare – architectures of present and future information technologies", *Clin. Chem. Lab. Med. (CCLM)*, vol. 57, no. 3, pp. 328-335, 2019.
[http://dx.doi.org/10.1515/cclm-2018-0658] [PMID: 30530878]

[11]   B. Mittelstadt, "Ethics of the health-related internet of things: A narrative review", *Ethics Inf. Technol.*, vol. 19, no. 3, pp. 157-175, 2017.
[http://dx.doi.org/10.1007/s10676-017-9426-4]

[12]   M.M. Psiha, and P. Vlamos, "IoT applications with 5G connectivity in medical tourism sector management: third-party service scenarios", *Adv. Exp. Med. Biol.*, vol. 989, pp. 141-154, 2017.
[PMID: 28971423]

[13]   S. Latif, J. Qadir, S. Farooq, and M. Imran, "How 5G wireless (and concomitant technologies) will revolutionize healthcare?", *Future Internet*, vol. 9, no. 4, pp. 93-102, 2017.
[http://dx.doi.org/10.3390/fi9040093]

[14]   F. O'Brolcháin, S. de Colle, and B. Gordijn, "The ethics of smart stadia: A stakeholder analysis of the Croke Park project", *Sci. Eng. Ethics*, vol. 25, no. 3, pp. 737-769, 2019.
[http://dx.doi.org/10.1007/s11948-018-0033-5] [PMID: 29497969]

[15]   P. Gope, and T. Hwang, "BSN-Care: A secure IoT-based modern healthcare system using body sensor network", *IEEE Sens. J.*, vol. 16, no. 5, pp. 1368-1376, 2016.
[http://dx.doi.org/10.1109/JSEN.2015.2502401]

[16]   V. Özdemir, and N. Hekim, "Birth of industry 5.0: making sense of big data with artificial intelligence, "The Internet of Things" and next-generation technology policy", *OMICS*, vol. 22, no. 1, pp. 65-76, 2018.
[http://dx.doi.org/10.1089/omi.2017.0194] [PMID: 29293405]

[17]   Li, D, "5G and intelligence medicine—how the next generation of wireless technology will reconstruct healthcare", *IEEE transaction*, vol. 2, pp. 205–208, 2019.

[18]   C.L. Russell, "5G wireless telecommunications expansion: Public health and environmental implications", *Environ. Res.*, vol. 165, pp. 484-495, 2018.
[http://dx.doi.org/10.1016/j.envres.2018.01.016] [PMID: 29655646]

[19]   P. R. Chai, H. Zhang, G. D. Jambaulikar, E. W. Boyer, L. Shrestha, L. Kitmitto, P. G. Wickner, H. Salmasian, and A. B. Landman, "An Internet of Things button to measure and respond to restroom cleanliness in a hospital setting: descriptive study," *Int. J. Adv. Comput. Sci. Appl.*, vol. 21, 2019.

[20]   G.B. Stefano, and R.M. Kream, "The Micro-Hospital: 5G telemedicine-based care", *Med. Sci. Monit. Basic Res.*, vol. 24, pp. 103-104, 2018.
[http://dx.doi.org/10.12659/MSMBR.911436] [PMID: 30006501]

[21]   G.J. Joyia, R.M. Liaqat, A. Farooq, and S. Rehman, "Internet of Medical Things (IoMT): Applications, benefits and future challenges in healthcare domain", *JCM,* pp. 240-247, 2017.

[22]   D. Gruson, "Ethics and artificial intelligence in healthcare, towards positive regulation", *Soins,* vol. 64, no. 832, pp. 54-57, 2019.
[http://dx.doi.org/10.1016/j.soin.2018.12.015] [PMID: 30771853]

[23]   K. Nikus, J. Lähteenmäki, P. Lehto, and M. Eskola, "The role of continuous monitoring in a 24/7 telecardiology consultation service—a feasibility study", *J. Electrocardiol.,* vol. 42, no. 6, pp. 473-480, 2009.
[http://dx.doi.org/10.1016/j.jelectrocard.2009.07.005] [PMID: 19698956]

[24]   S.B. Baker, W. Xiang, and I. Atkinson, "Internet of Things for smart healthcare: Technologies, challenges and opportunities", *IEEE Access,* vol. 5, pp. 26521-26544, 2017.
[http://dx.doi.org/10.1109/ACCESS.2017.2775180]

[25]   S. Tyagi, A. Agarwal, and P. Maheshwari, " A conceptual framework for IoT-based healthcare system using cloud computing", *International Conference—Cloud System and Big Data Engineering, Conflu,* vol. 2, pp. 34-48, 2016.

CHAPTER 8

Cyber Security in Healthcare

Rohini Chavan[1,*] and **Shreyash Shabadi**[1]

¹ E & TC Department, Vishwakarme Institute Of Technology, Pune, India

Abstract: Cyber security in healthcare focuses on safeguarding electronic information and assets against unauthorized access, use, or disclosure. The healthcare sector faces significant cyber security risks, with patient care and safety hanging in the balance. Its size, reliance on technology, sensitive data, and susceptibility to disruptions make healthcare facilities appealing targets for cybercriminals. In recent years, cyber-attacks have surged across various sectors, such as healthcare, finance, and manufacturing. In particular, the healthcare industry has become a top target due to insufficient security measures, outdated practices, and valuable sensitive data such as usernames, passwords, and medical records.

Keywords: Cyber-attack, Cybercriminals, Health care, Medical records, Security, Security risks.

INTRODUCTION

Cyber-attacks are increasingly targeting the healthcare industry, making it a primary focus for attackers. Future studies will concentrate on the application of cyber security within the healthcare sector, particularly exploring the various techniques used to defend Internet of Things (IoT)-based healthcare systems [1]. The study also examines various types of security threats within the healthcare industry. Cyber security is a rapidly evolving field of research with diverse applications across multiple sectors, including government services, military, healthcare, education, and manufacturing [2]. Financial services and transportation are also sectors impacted by cyber security risks. Previously, financial services were considered the most at-risk, but by 2015, the healthcare sector became the leading target for cyber-attacks [3].

Many cyber security applications are based on the Internet of Things (IoT), where IoT applications present unique challenges in terms of data transmission methods,

* **Corresponding author Rohini Chavan:** E & TC Department, Vishwakarme Institute Of Technology, Pune, India; E-mail: rohini.chavan@viit.ac.in

Parikshit N. Mahalle, Gitanjali R. Shinde, Namrata N. Wasatkar & Prashant R. Anerao (Eds.)

posing various cyber security difficulties. The Internet of Things (IoT) connects cyberspace with the physical world. In IoT-based healthcare applications, multiple sources are used to collect patient data, which is then compiled into Electronic Health Records (EHR). These records can be either transferred over the internet or uploaded to the cloud. The Electronic Health Record (EHR) is a structured way to collect all electronic health information related to patients [4]. Additionally, these digital records can be shared across various healthcare settings [5]. These include various types of data such as demographics, medication history, medical conditions, and laboratory test results. Additionally, it contains the patient's billing information [6]. Healthcare applications are very critical and need to be powerfully protected.

Cyber security has not received sufficient attention in the healthcare sector, despite its critical importance for patient safety and hospital reputation. To prevent data breaches that could compromise patient privacy, hospitals must implement effective IT security measures. This research article reviews several professional publications addressing ransomware attacks and other cyber-attacks on hospitals from 2014 to 2020. The report highlights the latest defence strategies presented in the scientific literature that can be applied within the healthcare industry. Additionally, it provides an overview of the impacts of cyber-attacks and the steps hospitals have taken to manage and recover from these incidents [7, 8]. The study emphasizes that cyber-attacks on hospitals have serious consequences and underscores the need for prioritizing cyber security in healthcare. To counter cyber-attacks, hospitals must establish clear policies and backup plans, regularly update their systems, and train staff to identify and handle online threats. The paper concludes that implementing robust cyber security measures can mitigate the damage caused by system failures, reputational harm, and other related issues [9].

The intelligent Internet of Things (IoT), which provides an endless amount of networking possibilities for exploring medical data, is improving the connection between technology and healthcare society [10]. Deep networks have seen productive changes in recent years, and the use of medical wearables has become more widespread. The Internet of Things enabled by deep neural networks has led to innovative social breakthroughs in medicine and opened new avenues for health data research [11]. So these kinds of applications do not use old-style safety appliances. However, not all the mechanisms are entirely newly produced. Some traditional mechanisms are used with slight modifications in it. Thus the healthcare industries have to be muscularly secured and protected from any kind of attacks and cyber-criminals. This chapter aims to assess the most common threats that face the healthcare industries and their countermeasures [12].

Despite the progress that has been made, there are still several issues that need to be addressed in terms of quality of service. Gray Filter Bayesian Convolutional Neural Network or GFB-CNN is a smart healthcare method that is powered by Deep Neural Networks and uses real-time data [13]. This technique is presented here as part of this study.

The health zone can have a lot of security concerns, especially with the Internet, where cyber-attacks are becoming more common and sophisticated. Rights of entry and control breaches, attacks that breach and run malware, and DoS (denial of carrier) attacks are some of the maximum common threats to healthcare protection [14]. When evaluating paid Denial of carrier (DDoS) attacks that lease multiple hosts to attack a machine, DoS attacks involve a free pool that floods the target machine with requests. This makes it difficult to determine the origin of the attack. Patients may also suffer as a result of these attacks, and healthcare agencies may additionally suffer from reputational damage [15]. The next most important threat to the healthcare sector is malware, which comes in new forms all the time. Ransomware is one family of malware that healthcare facilities are becoming interested in. Ransomware was ranked 2nd on the list of cyber security risks for healthcare businesses in a survey with the help of the Healthcare Statistics and Management Structure Society (HIMSS), with 17% of respondents saying they had been the victim of a ransomware attack [16]. Healthcare businesses have been the target of several high-profile cyber-attacks in recent years, for example, the 2021 ransomware attack that hit the Irish government fitness carrier (HSE) severely disrupted healthcare services [17]. Similarly, more than 150 countries were affected by the Wanna-Cry ransomware attack in 2017, which prompted the UK's National Fitness Provider (NHS) to reschedule tactics and cancel appointments. Healthcare facilities should put robust cyber safeguards in place to stop cyber-attacks and protect sensitive patient information [18].

However, many healthcare facilities lack adequate security protocols, leaving them open to intrusions. The simplest 44% of healthcare agencies, in line with the view through the Ponemon Institute, have intensive protection around based on responses from 167 healthcare cyber security specialists. The author's approach concerned conducting a content evaluation of clinical papers from 2014 to 2020 that mentioned malware, DoS, and social engineering attacks on hospitals [19]. There are five sections in the document. Hospitals that have carried out qualified cyber-attacks between 2014 and 2020 are classified as Phase II. Hospitals can practice the measures discussed in Phase III to mitigate or prevent a cyber-attack. Effects and discussions are presented in Part IV, and the paper is wrapped up in Phase V.

CYBERSECURITY IN HEALTHCARE SYSTEMS

This section covers several hospitals that faced cyber-attacks, how we handled the situations, and the impact of the attacks. Table **1** shows the ways hackers target the healthcare sector and the steps hospitals take to prevent and recover from these attacks.

Table 1. Methods of cyber-attack on hospitals and responses to cyberattacks.

Sr. No.	Hospital	Attack Methods	Response	Source
1	Boston Children's Hospital	Hackers tried to compromise the hospital network by concentrating on "exposed ports and services," as well as by sending out a phishing email campaign that was directed at hospital staff members.	To properly seal all firewall entry points and protect employees from inadvertently clicking on a dangerous link, the hospital took the precaution of shutting down all web-facing applications, including email services.	[20]
2	Lukas Hospital	Technique for social engineering.	Every system has been shut down. Systems were restored using backups.	[21]
3	Hancock regional hospital	The hackers exploited the Microsoft Remote Desktop Protocol to gain access to a hardware vendor's administrative account.	Shut down all desktop and network systems.	[21]
4	Hollywood Presbyterian Medical	NAN	Pay a ransom	[22]
5	Brno University Hospital	Taking advantage of a vulnerability in the Windows XP operating system.	Shut down the entire information technology network.	[23]

All these hospitals, including Boston Hospital, Lukas Hospital, Brno Hospital, and Hancock Hospital, were hit by different cyber-attacks and responded in the same way: they shut down their systems to minimize damage. The table shows that these hospitals lacked clear strategies or backup plans to handle such attacks, highlighting a significant gap in preparedness. For example, Brno Hospital was still using Windows XP in 2020. This shows how crucial it is for healthcare organizations to focus on cyber security and take steps to reduce and prevent online threats. Cyber-attacks can be grouped into three main types:

1. **Injection attack:** An attacker can "inject" malicious data into a web application, changing how it works by making it run certain commands. Injection is one of the first types of web attacks. Malware is an example of an

injection attack. According to a study [20], malware is any computer code designed to gain unauthorized access to digital devices and IT systems by bypassing security measures and taking advantage of weaknesses. Three main types of malware were identified:

   a. **SamSam:** SamSam is a type of ransomware that first appeared in late 2015 and is primarily targeted at the healthcare industry. It specializes in exploiting vulnerabilities in RDP (Remote Desktop Protocol), FTP (File Transfer Protocol), and Java-based web servers to gain unauthorized access to victim systems.

   b. **Locky:** Locky is a ransomware family introduced in 2016 that uses a hybrid encryption system. It operates by scanning the victim's drives, including network drives, for specific file types to encrypt. The files are then encrypted using both RSA and AES encryption methods.

   c. **Net walker:** Also known as Mailto, Net walker is a ransomware attack that encrypts all Windows devices within a victim's network. The attack can be initiated through phishing emails or executable files that spread throughout the network to compromise systems.

2. **Social Engineering:** Social engineering is a technique where attackers manipulate individuals into revealing sensitive information by exploiting their psychological vulnerabilities. Phishing is a common form of social engineering, where attackers trick victims into providing private details such as usernames, passwords, and bank account information. This is usually done by luring victims into clicking on links to fraudulent websites or downloading malicious software.

3. **Denial of Service (DoS) Attack:** This type of cyber-attack primarily aims to use up resources, such as memory or processing power. It can be carried out over both wireless and wired connections. A specific form of DoS attack called a distributed denial-of-service (DDoS) attack, targets websites. In this case, the attacker uses a malicious script installed on multiple computers to attack a single victim, causing the website to go offline [21].

MITIGATION STRATEGIES

A web application may be "injected" with malicious data by an attacker, affecting the way it operates by directing it to execute certain commands. Injection is one of the earliest web-based attacks. An injection attack is depicted by a piece of software. The purpose of the computer code is to gain unauthorized access to digital devices and IT infrastructures. This is done by taking advantage of security flaws and breaching the security measures protecting them. Three different types of malware were present.

SamSam was first appearing in late 2015, it was primarily targeted at the healthcare sector. SamSam uses web server vulnerabilities to access the victims' machines.

Locky is a ransomware family that employs a hybrid coin system. Its operation involves scanning the victim's drives, including network drives, for specific file types to encrypt and protect.

Mailto is a type of attack in which the attacker leverages the victim's network to compromise their Windows-based devices. The attacker may use various methods to carry out the attack.

Social engineering is a technique where an attacker interacts with the victim to manipulate them into providing sensitive information.

According to a study [13], risk refers to the potential for loss or damage if a security vulnerability is exploited by an attacker. Another broad definition of cyber security risk involves the operational risks from online activities that threaten information assets, IT resources, and technology, which can result in damage to both tangible and intangible assets, business disruptions, and harm to an organization's reputation [22]. Risk mitigation strategies offer two options: risk reduction and risk avoidance. These strategies use preventive measures to minimize the likelihood or impact of a cyber-attack. Their focus is on identifying and addressing any weaknesses or security risks within an organization's policies and information. Risk mitigation tactics can include installing intrusion detection systems, building protective barriers, regularly updating software and hardware, and training staff on cyber security best practices.

Proactive Incident Response (IR)

The process consists of six steps: planning and preparation, detection, analysis and evaluation, containment and eradication, recovery, and post-incident activities. A business must begin by creating its security policy and setting up its incident response capability. This involves forming an incident management team and gathering the necessary tools and resources. In the second phase, the incident is detected automatically using tools like network or host-based intrusion detection systems, or manually through methods such as user reports of issues. In the third phase, the incident response team analyses and verifies the incident. During the fourth phase, containment measures like sandboxing are put into place. In the fifth phase, the administrator ensures that systems are functioning properly and resolves any issues to prevent future incidents. The final step is a post-incident meeting, aimed at improving technology and gaining insights [22].

Secure Architecture Based on Block-chain Technology and Artificial Intelligence

The proposed architecture for a secure system utilizing artificial intelligence and blockchain technology consists of five layers. The first layer, called the "data layer," gathers information from patient sensors, such as temperature and heart rate, and also collects malware samples that are sent to the next layer, the malware analysis layer. In the second layer, malware analysis tools like Pseudo and Process Explorer are used to examine the malware. In the third layer, called the intelligence layer, harmless malware samples from the second layer are checked for security vulnerabilities using AI techniques such as support vector machines (SVM) and random forests (RF). Data from the third layer is securely stored in the blockchain in the fourth layer. The application layer (layer five) includes recipients of the health data, such as hospitals, pharmacies, laboratories, and ambulances.

Design of a Multi-agent Framework

The framework is developed in two stages. First, five system agents must be created: Patient, Nurse, Doctor, Ambient, and Database Agents. The second step is to design a layered architecture that organizes the agents based on their data storage and power capabilities. In this framework, a wireless sensor network platform was used [22].

Scheme Relies on Stacked Autoencoder for Intrusion Detection

The scheme intrusion detection framework employs stacked autoencoders. The method consists of three steps: data pre-processing, feature extraction, and intrusion behaviour detection. During the data pre-processing phase, infiltration behaviour is defined. In the feature extraction phase, a stacked autoencoder is used to extract parameter weights for various features. Finally, in the intrusion behaviour detection phase, the Boost algorithm is applied to determine whether the behaviour is normal or intrusive.

SECURITY THREAT IN HEALTHCARE INDUSTRY

Healthcare applications are highly sensitive, and medical data requires more robust security measures than other types of data and applications. Various security threats specifically impact healthcare applications, each with unique causes and solutions. This heading covers some of these security threats, including eavesdropping, impersonation, message modification, and man-in-t-e-middle attacks.

- **Eavesdropping Attack:** In this scenario, Alice, Bob, and the Attacker are the actors. An eavesdropping attack happens when an attacker intercepts all the data transmitted from the sender to the receiver. In healthcare applications, this can compromise the privacy of medical data. For example, if Alice sends unencrypted data to Bob, an attacker can use sniffing tools to intercept and capture the communication.
- **Impersonation:** Impersonation involves an attacker taking on the identity of an authorized user and transmitting data through the user's IP address without undergoing any authentication process.
- **Message Modification:** Message modification works as follows: Alice sends data from the transmitter side, and the attacker intercepts this data. The attacker then alters the content of the message and resends it to the receiver, Bob, using Alice's IP address.
- **Man-In-The-Middle:** A man-in-the-middle attack occurs when an attacker intercepts the communication between a legitimate sender and receiver, allowing them to modify the messages exchanged between the two without either party being aware of the alteration.

To protect healthcare applications from various security attacks, different security mechanisms must be implemented. For instance, by using a robust authentication system and a strong encryption and decryption model for medical data, we can effectively secure the data and prevent it from being attacked.

SECURITY COUNTERMEASURES

To address security threats, various countermeasures are used, with different solutions implemented to handle specific types of threats. Some of these solutions include Encryption, Authentication, and Authorization, all aimed at protecting stored data from various security risks. This heading focuses on encryption algorithms as a means to safeguard data from different threats, though not all cases are covered. In some instances, both encryption and authentication methods are required to protect data, especially in the IoT framework, which is dynamic. However, traditional security countermeasures cannot be directly applied to IoT applications; instead, they serve as the foundation for new methods. This paper discusses some of the most common symmetric encryption algorithms, including DES, 3DES, Blowfish, and AES.

- **DES**: (Data Encryption Standard) is an encryption algorithm developed by IBM in 1977. It is designed to encrypt data by processing a fixed-length stream of plaintext bits, converting them into cipher text of the same size. The block size is 64 bits, with 56 bits used for the encryption key and the remaining 8 bits reserved for error checking. However, DES is considered a relatively slow encryption algorithm [23].

- **Bluefish:** Developed in 1993, the Blowfish algorithm uses a block size of 64 bits and a key size ranging from 32 to 448 bits. It was introduced as an alternative to the DES algorithm, offering improved speed and security with a variable key size. Blowfish is available under a free license and can be used by anyone. It is generally considered faster and more secure than both DES and 3DES [23].
- **3DES:** Triple Data Encryption Standard, introduced in 1998 as an improvement of DES. The algorithm uses DES three times. It also uses the same number of 64 elements as the standard size of 56 elements. This algorithm is considered faster than DES but is also considered slower because it uses DES three times. It is safer than DES [23].
- **Advanced Encryption Standard (AES):** Developed in 2001 by the National Institute of Standards and Technology (NIST), AES uses a 128-bit block size with key sizes of 128, 192, or 256 bits. The number of rounds varies depending on the key size: 10 rounds for 128-bit keys, 12 rounds for 192-bit keys, and 14 rounds for 256-bit keys. AES is considered a fast and secure encryption algorithm [24].

RESULTS AND DISCUSSION

As healthcare organizations become more attractive targets for hackers, cyber-attacks on hospitals are increasing, as highlighted in a review of scholarly articles for this research paper. The healthcare sector is particularly vulnerable due to outdated practices, weak security measures, and the sensitivity of the data it handles, making it a prime target for cybercriminals. These attacks can lead to severe consequences, such as harm to patients, damage to the reputation of healthcare organizations, and financial losses.

The study also highlights various protective measures against cyber-attacks and solutions that healthcare institutions can implement. These include employee training, regular system updates, and the use of advanced security tools like intrusion detection systems and firewalls. The research suggests that hospitals should prioritize cyber security and establish clear strategies and backup plans for responding to breaches.

Expert studies indicate that hospitals are particularly susceptible to attacks due to insufficient security standards. Many healthcare organizations lack a comprehensive security strategy, reflecting a neglect of cyber security. The study recommends that healthcare facilities take proactive steps to safeguard sensitive patient data and reduce the impact of system failures, reputational harm, and other related issues.

CONCLUSION AND FUTURE WORKS

This study concludes by emphasizing the critical need for healthcare facilities to address cyber security in order to prevent data breaches that could jeopardize patient information. Cyber-attacks on hospitals can have severe consequences, so healthcare organizations must establish clear policies and contingency plans to manage these incidents. The report provides an overview of hospital cyber-attacks from 2014 to 2020, including ransomware attacks, and suggests various strategies hospitals can use to reduce or prevent hacking.

Future research could focus on developing innovative techniques and tools to protect healthcare organizations from cyber-attacks. For example, studies could explore how machine learning and artificial intelligence can be used to detect and prevent cyber-attacks in hospitals. Additionally, future research might examine the impact of cyber-attacks on patient safety and explore the ethical considerations of data breaches in healthcare.

Overall, this study underscores the importance of cyber security in the healthcare sector and stresses the need for healthcare organizations to take proactive measures to protect sensitive patient data. Given the rise in cyber-attacks on hospitals, healthcare institutions must prioritize cyber security to minimize losses due to system failures, reputational damage, and other related challenges.

REFERENCES

[1]   J. Mirkovic, and P. Reiher, "A taxonomy of DDoS attack and DDoS defense mechanisms", *Comput. Commun. Rev.,* vol. 34, no. 2, pp. 39-53, 2004.
       [http://dx.doi.org/10.1145/997150.997156]

[2]   M. O'Brien, "Ireland's Health Service Executive hit by ransom ware attack", *IEEE Conference*, pp. 12-17, 2023.

[3]   A. Osborn, "NHS cyber-attack: GPs and hospitals hit by ransomware," *IEEE Transaction*, pp. 103-112, 2023.

[4]   Ponemon Institute, "Sixth annual benchmark study on privacy & security of healthcare data", *Proceedings of the 11th international conference on cyber warfare and security,* pp. 23-40, 2023.

[5]   M. Ashawa, and T. Morris, "Understanding and mitigating malware attacks", *Proceedings of the 11th International Conference on Cyber Warfare and Security (ICCWS 2019),* vol. 1, pp. 1-10, 2019.

[6]   V. Arora, A. Varshney, A. Arora, and N. Shukla, "Assessment of SamSam Ransomware Attack on Healthcare Sector and Way Forward", *J. Inf. Priv. Secur.,* vol. 15, no. 1, pp. 1-12, 2019.

[7]   S. Almashhadani, T. Almarshad, and A. Al-Salman, "Ransomware: The past, present, and future", *Proceedings of the 3rd International Conference on Computer Applications & Information Security (ICCAIS),* vol. 1, pp. 1-6, 2019.

[8]   J.A. Gómez-Hernández, P. García-Teodoro, and J.E. Díaz-Verdejo, "Analysis of netwalker

ransomware: Detection, prevention and recovery", *Comput. Secur.,* pp. 106-120, 2023.

[9]   W. Wang, Y. Zeng, X. Zhang, X. Xu, Y. Xiang, and X. Shen, "A survey on social engineering attacks and defenses in online social networks", *IEEE Commun. Surv. Tutor.,* vol. 22, no. 2, pp. 1342-1372, .

[10] R. Singh, S. Singh, and D. Saini, "Denial of service attacks: Impact, detection, and mitigation techniques", *J. Netw. Comput. Appl.,* vol. 135, pp. 62-80, 2019.

[11] A. Singh, A. Kumar, and S. Tyagi, "A comparative analysis of detection and mitigation techniques against distributed denial of service attacks", *Proceedings of the International Conference on Smart Technologies in Computing and Communication,* pp. 259-269, .

[12] P. Kandasamy, M. Perumal, and R. Naresh, "Cyber-security risks and their mitigation strategies for healthcare industry", *Cyber-security and Privacy Issues in Industry 4.0.,* Singapore, Springer, pp. 19-37, 2022.

[13] A. S. Czewski, "Cyber-security risk management in the healthcare industry.", *IEEE Transactions,* pp. 103-116, 2020.

[14] Y. He, X. Lu, Y. Yao, W. Zhang, and W. Tang, "A cyber security incident response system with automated forensics and orchestration", *IEEE Access,* vol. 10, pp. 113773-113786, 2022.

[15] A. Alabdulatif, A. Ahmad, M.K. Khan, A. Azeem, A. Al-Khateeb, and A. Al-Salman, "A secure architecture based on block chain technology and artificial intelligence for healthcare applications", *Future Gener. Comput. Syst.,* vol. 127, pp. 487-495, 2021.

[16] R.A. Ramadan, B.W. Aboshosha, J.S. Alshudukhi, A.J. Alzahrani, A. El-Sayed, and M.M. Dessouky, "Cybersecurity and Countermeasures at the Time of Pandemic", *J. Adv. Transp.,* vol. 2021, pp. 1-19, 2021.
      [http://dx.doi.org/10.1155/2021/6627264]

[17] A. S. Wilner, H. Luce, E. Ouellet, O. Williams, and N. Costa, "Cyber-security and Canada's healthcare sector", *International Journal: Canada's Journal of Global Policy Analysis,* vol. 76, no. 4, pp. 522-543, 2021.

[18] A.K. Alharam, and W. El-madany, "The Effects of Cyber-Security on Healthcare Industry", *9th IEEE-GCC Conference and Exhibition (GCCCE),* 2017.
      [http://dx.doi.org/10.1109/IEEEGCC.2017.8448206]

[19] Paul III, D. P., Spence, N., Bhardwa, N., & PH, C. "Healthcare facilities: Another target for ransom ware attacks", *IEEE Conference Singapore,* pp. 1-15, 2021.

[20] L. Coventry, and D. Branley, "Cyber-security in healthcare: A narrative review of trends, threats and ways forward", *Maturitas,* vol. 11, pp. 48-52, 2022.

[21] H. Saleous, M. Ismail, S.H. AlDaajeh, N. Madathil, S. Alrabaee, K-K.R. Choo, and N. Al-Qirim, "COVID-19 pandemic and the cyber-threat landscape: Research challenges and opportunities", *Digit. Commun. Netw.,* vol. 3, pp. 1145-1159, 2022.

[22] H. Ghayoomi, K. Laskey, E. Miller-Hooks, C. Hooks, and M. Tariverdi, "Assessing resilience of hospitals to cyber-attack" *IEEE Transaction, Dubai,* vol. 7, pp. 12-23, 2022.

[23] D.N. Burrell, A.S. Aridi, Q. McLester, A. Shufutinsky, C. Nobles, M. Dawson, and S.R. Muller, "Exploring system thinking leadership approaches to the healthcare cybersecurity environment", *Int. J. Extrem. Autom. Connect. Healthc.,* vol. 3, no. 2, pp. 20-32, 2021. [IJEACH].
      [http://dx.doi.org/10.4018/IJEACH.2021070103]

[24] Strasburg, J., & Hinshaw, D., "Cyber-criminals Sweep In to Take Advantage of Coronavirus", *The Wall Street Journal,* vol. 24, pp. 45-55, 2020.

CHAPTER 9

An IoT-enabled Women's Security Device Utilizing Arduino for GPS Tracking and Alerts in the Context of Industry 5.0

Pallavi Devendra Deshpande[1,*]

[1] *Department of Electronics and Telecommunication Engineering, Vishwakarma Institute of Technology, Pune, India*

Abstract: India is becoming a superpower in today's fast-paced world thanks to its technological advancements and other areas. However, there has not been a significant decline in the crime rate against women and children. For this reason, to lower the number of crimes against women and children, greater awareness and technological assistance are required. Our proposal is for a "Safety Device using IoT" that tracks the victim's whereabouts continually using a GPS tracking system, calls phone numbers registered with the system, and sends out continuous SMS messages with the victim's location. We are attempting to address the current need for widely available, affordably accessible technology with this system that we are developing. The abstract presents a women's security device that enhances women's security through GPS tracking and alerts. The Arduino microcontroller platform serves as the system's main building block and allows for the integration of several sensors and communication tools. Women may always have it with them because the device is designed to be easily carried and portable. GPS tracking technology allows for real-time position monitoring, which can expedite emergency response times. The alert system uses a combination of vibration and sound to warn the user and anybody nearby in the event of danger.

Keywords: Current, Electronic appliances, Real-time monitoring, Health, Voltage.

INTRODUCTION

In the contemporary landscape, ensuring personal safety, especially for women, has become a paramount concern. The apprehension associated with traveling alone, particularly at night, underscores the need for innovative solutions that leverage technology to mitigate risks and provide a sense of security. Women, who are often perceived as more vulnerable to various forms of violence, inclu-

* **Corresponding author Pallavi Devendra Deshpande:** Department of Electronics and Telecommunication Engineering, Vishwakarma Institute of Technology, Pune, India; E-mail: pallavi.deshpande@viit.ac.in

Parikshit N. Mahalle, Gitanjali R. Shinde, Namrata N. Wasatkar & Prashant R. Anerao (Eds.)

ding robbery, sexual assault, rape, and domestic violence, can benefit significantly from advancements in personal safety devices. The prevailing societal challenges have prompted a reevaluation of safety measures, acknowledging the need for proactive strategies to reduce the likelihood of individuals, particularly women, becoming victims of violent crimes. Recognizing and responding to unsafe situations is crucial, and this has spurred the development of technology-driven solutions aimed at empowering individuals to enhance their personal safety. One notable technological intervention in this domain is the integration of GPS tracking and alerts in personal safety devices. This advancement serves as a beacon of hope, offering women a tool to bolster their confidence when traversing unfamiliar or potentially risky environments, such as walking alone or commuting. The convergence of technology, in this case, revolves around the implementation of the Arduino platform, a versatile microcontroller system renowned for its adaptability to integrate various sensors and communication technologies. The women's safety device utilizing GPS tracking and alerts, based on the Arduino platform, epitomizes the fusion of hardware and software to create a portable, user-friendly solution. Arduino's programmable nature enables the incorporation of diverse sensors, including GPS modules, accelerometers, and communication interfaces, to craft a comprehensive safety apparatus. The GPS tracking feature allows individuals, and specifically women in this context, to share their real-time location with trusted contacts or emergency services. This functionality proves invaluable in situations where immediate assistance is required. Additionally, the device can be programmed to send alerts or distress signals if predefined parameters indicative of potential danger are met. The portable nature of this safety device ensures ease of use, enabling women to carry it effortlessly during their daily activities. Its discreet design contributes to user comfort, fostering a sense of empowerment rather than intrusion. Moreover, the integration of Arduino facilitates customization, allowing for the adaptation of the device to cater to specific user preferences and requirements. The impact of such technological innovations extends beyond the individual level. By providing women with tools that enhance their safety and security, society takes a collective step toward creating an environment where everyone feels protected. The synergy between personal safety devices and technology exemplifies a progressive approach to addressing societal challenges, reaffirming the potential of innovation to contribute to the well-being and empowerment of individuals, particularly women, in their daily lives. In conclusion, the incorporation of GPS tracking and alerts in personal safety devices utilizing Arduino technology represents a commendable stride towards fostering women's safety. As technology continues to evolve, the synergy between innovation and personal security endeavors to create a world where individuals, irrespective of gender, can navigate their surroundings with confidence and resilience against potential threats [1 - 3].

In an era where personal safety is a paramount concern, especially for vulnerable individuals such as children and women, technological innovations play a crucial role in providing effective solutions. This book chapter explores the intricacies of the "Safety Device using IoT," a comprehensive system designed to address emergency situations and enhance the security of individuals in distress. Comprising key components such as the Arduino Nano, NEO6M GPS module, GSM technology, SOS button, RF transmitter, and RF receiver, this safety device represents a cutting-edge application of the Internet of Things (IoT) in ensuring personal safety [4 - 6].

Components Overview

Arduino Nano with Atmega 328P Microcontroller

It serves as the central processing unit, facilitating the integration and coordination of various components, and enables seamless communication between different modules and sensors.

NEO6M GPS Module

It provides accurate global positioning system (GPS) data for precise location tracking and enables essential real-time monitoring of the victim's location during distress.

GSM (Global System for Mobile Communication)

It facilitates communication by sending messages and making calls to pre-registered phone numbers and enables a reliable and immediate alert mechanism.

SOS Button

It functions as the trigger for activating the safety device in emergencies and is designed for easy operability, ensuring quick response during distress.

RF Transmitter and RF Receiver

It enables seamless communication within the system, enhancing the overall reliability of the safety device and contributing to the comprehensive communication infrastructure of the device. Fig. (**1**) shows Architectural overview of Women's security devices with IoT connectivity.

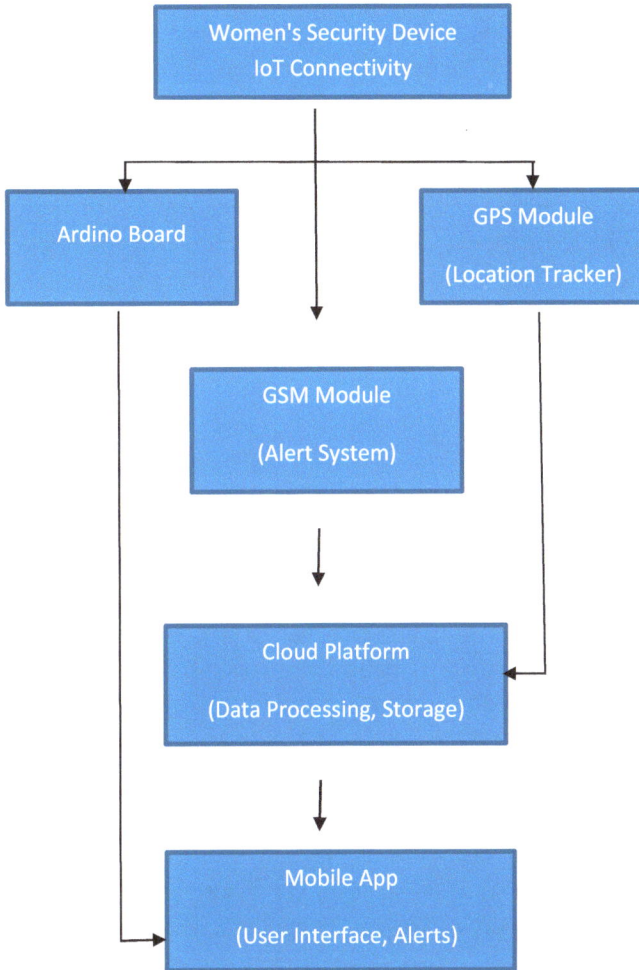

Fig. (1). Architectural overview of women's security devices with IoT connectivity.

Operational Workflow

When an individual, whether a child, woman, or any person, finds themselves in distress, the activation of the SOS button initiates a series of crucial actions. The safety device promptly sends messages to pre-registered phone numbers, which can include parents, friends, police stations, or other designated guardians. This multi-contact approach ensures that a broader network is alerted, increasing the likelihood of timely assistance. Simultaneously, the safety device makes calls to the registered phone numbers, further enhancing the chances of immediate attention. The device's integration of continuous SMS alerts and phone calls ensures redundancy in the communication mechanism, increasing the reliability of the emergency alert system. A standout feature of the safety device is its ability to track the victim's live location continuously. Leveraging the NEO6M GPS module, the device provides real-time location data, offering a valuable tool for authorities and contacts to assess the situation accurately. This live tracking feature significantly enhances the efficiency of emergency response, enabling prompt action based on precise location information. The operational design of the safety device prioritizes easy operability, ensuring that individuals, including children or those in distress, can easily activate the device during emergency situations. The straightforward activation process through the SOS button contributes to the user-friendly interface, allowing for a quick response in critical moments. Furthermore, the safety device places a strong emphasis on cost efficiency to make this crucial safety solution accessible to a broad audience. The overarching goal is to reach the maximum number of people and provide them with a reliable and affordable service. The integration of cost-effective components and the streamlined design aligns with the objective of ensuring widespread accessibility. The "Safety Device using IoT" represents a significant leap in the realm of personal safety, leveraging IoT technologies to create a robust and versatile solution. With a focus on ease of use, comprehensive functionality, and cost efficiency, this safety device offers a beacon of hope for individuals in distress. As technology continues to evolve, IoT-based safety devices stand at the forefront of innovative solutions, providing a sense of security and empowering individuals to navigate emergency situations with confidence.

LITERATURE SURVEY

The integration of a GSM module and an Arduino Uno microcontroller is a pivotal aspect of the safety device, combining hardware and software elements to ensure seamless connectivity and functionality. The GSM module relies on Global System for Mobile Communications (GSMTM) technology, a widely used standard that employs Time Division Multiple Access (TDMA) signaling over Frequency Division Duplex (FDD) carriers with Phase Shift Keying (PSK) modulation. This technology facilitates efficient communication and data transfer, essential for the effective operation of the safety device. The choice of GSMTM technology aligns with the goal of creating a cost-effective solution, ensuring that the device remains affordable for the typical Indian user. By incorporating components that maintain a balance between functionality and production costs, the safety device becomes accessible to a broader demographic. The emphasis on cost-effectiveness is particularly crucial in addressing the needs of users in India, where affordability is a key consideration. In the context of women's safety, this paper makes a significant contribution by proposing a tool that can aid Indian women in distress. The device's design prioritizes cost efficiency to cater to the economic realities of the typical Indian consumer. This commitment to affordability enhances the device's potential impact, ensuring that it can reach and benefit a larger segment of the population. Overall, the integration of the GSM module and Arduino Uno microcontroller underscores the device's technological foundation, which, when combined with a focus on cost-effectiveness, makes strides in advancing the field of women's safety in India [7 - 9].

The paper introduces a novel and practical strategy to address the issue of crime against women in India by proposing a low-cost safety device. One of the key features of this innovative approach is the utilization of SMS technology to transmit the location's coordinates in case of an emergency. This method serves as a proactive and immediate response mechanism, aiming to enhance the safety of Indian women who may find themselves at risk. The safety gadget, as recommended in the paper, offers a straightforward yet effective solution. In the event of an emergency, the device is designed to send an SMS containing the precise location coordinates to pre-programmed mobile numbers. This feature ensures that designated individuals, such as parents, friends, or authorities, receive immediate information about the user's location when an emergency situation arises. The proposed safety device stands out for its cost-effectiveness, aligning with the economic realities of the typical Indian consumer. By incorporating this technology into a low-cost solution, the paper suggests a practical means of enhancing women's safety in India. The utilization of SMS for location sharing adds a layer of simplicity and accessibility to the device, making it user-friendly and applicable in various emergency situations. In conclusion, the paper's recommendation for a low-cost safety device with SMS-based location sharing presents a viable and impactful strategy for addressing crime against women in India. The emphasis on practicality, cost-effectiveness, and immediate response contributes to the device's potential effectiveness in enhancing the safety and security of women facing potential risks [10 - 12].

The survey delves into the realm of women's safety devices that leverage the Internet of Things (IoT) technology. These safety devices are specifically crafted to cater to women facing dangerous or emergency situations, emphasizing simplicity, portability, and multi-functionality. Recognizing the need for innovative solutions, both governmental bodies and individuals have contributed to the creation of various mobile applications and smart devices aimed at enhancing women's safety. The survey underscores the challenges associated with relying solely on mobile phones during emergencies, acknowledging the limitations that may arise in critical situations. To address these concerns, women's safety devices utilize a range of techniques, including location tracking, notifications, sensors, and image capture functionalities. While the literature reveals the existence of auto-detection women's safety devices based on parameters like voice recognition, temperature, and heart rate, the survey highlights potential shortcomings in such systems, particularly in cases of women with abnormal health conditions. The core functionality of these devices involves determining the user's location through the integration of GPS and GSM technology. Once the location is identified, the system communicates this information to designated individuals through various channels such as SMS, email, and phone calls. An audible alarm, generated by a buzzer, further enhances the user's ability to attract attention and seek assistance from people in close proximity. The survey recognizes the significance of sensors in these safety devices, describing them as devices that calculate or detect physical properties, providing crucial information or triggering responses in tandem with other devices. This holistic exploration of women's safety devices using IoT sheds light on the diverse technological approaches employed to ensure the well-being of women in various situations.

This study explores multiple safety devices designed to ensure women's security, incorporating advanced technologies to provide effective security measures. Each system offers unique features and functionalities, contributing to the overall landscape of women's safety solutions [13 - 15].

GPS-Equipped Smart Watch with Voice Recognition

One proposed solution involves a GPS-equipped smartwatch equipped with voice recognition technology, an electric shock-generating module, and screaming alarm modules. The smartwatch incorporates three sensors—a temperature sensor, a pulse rate sensor, and a motion sensor. This comprehensive system enables real-time monitoring and can identify when a woman is in a dangerous scenario. By leveraging technology for location tracking and health monitoring, the system facilitates prompt and appropriate action to ensure women's safety.

Security System with Multiple Modules

Another recommendation focuses on a security system that integrates various modules for robust safety measures. The system comprises the GSM shield (SIM900A), Atmega328 board, Arduino board, GPS (GYGPS6MV2) module, screaming alarm (ADR 9600), pressure sensor, and power supply unit. By combining these components, the system aims to provide comprehensive security precautions for women. The diverse set of modules enhances the system's capabilities, offering a multifaceted approach to women's safety.

"FEMME" Device with Bluetooth Synchronization

The "FEMME" device introduces a unique approach by utilizing an ARM controller and synchronizing with a smartphone *via* Bluetooth. This device can be triggered independently and records audio. It also features a hidden camera detector, providing additional layers of security. The user-friendly design enhances accessibility, making it a practical option for women seeking reliable safety solutions.

"SURAKSHA" Device with Voice Activation and Force Sensors

The "SURAKSHA" device employs three activation methods—voice, switch, and shock. Notably, the device automatically locks when not in use to prevent unnecessary signals. It can be activated by a voice command, force sensors that trigger when thrown by force, or a simple press of the switch during distress. The combination of these activation methods ensures versatility and responsiveness in various situations.

Wearable Sensor Nodes with Solar Energy Harvesting

This study introduces wearable sensor nodes equipped with solar energy harvesting capabilities. The system incorporates multiple sensors to monitor an individual's health data. Additionally, a single online application is developed to track the data collected by sensor nodes. The integration of solar energy harvesting adds sustainability to the wearable system, ensuring prolonged functionality without frequent battery replacements. In conclusion, these innovative safety devices showcase the diverse approaches and technological advancements in ensuring women's safety. From smartwatches with comprehensive sensor arrays to devices with voice activation and solar-powered wearables, these solutions contribute to creating a safer environment for women. The multi-functionality, integration of advanced technologies, and user-friendly designs collectively represent a significant stride toward addressing the security concerns of women in various contexts.

SYSTEM ARCHITECTURE

There are 2 segments of the system, the transmitter end and the receiver end. Block diagrams of the transmitter end and receiver end are shown in Figs. (**2** and **3**) as follows:

GPS Module

This module will be used to determine the user's location. Through serial connectivity, the GPS module will talk to the Arduino.

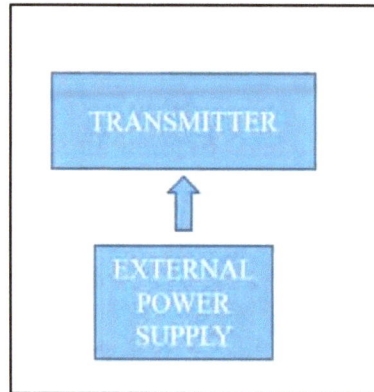

Fig. (2). A diagram of a transmitter with its power supply.

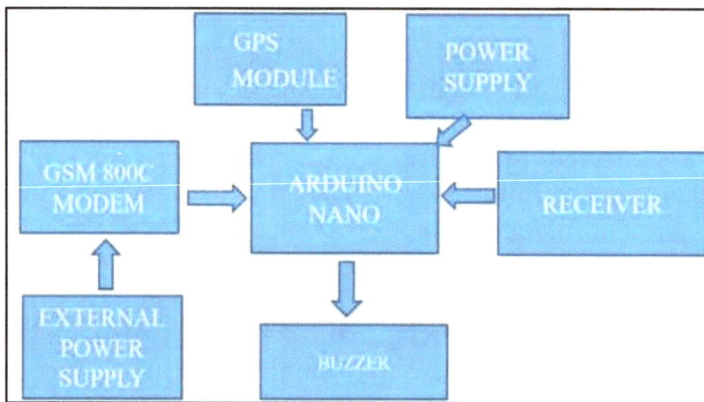

Fig. (3). A diagram of a receiver system with a GPS module, Arduino nano, and buzzer.

Arduino Board

The Arduino board will serve as the system's primary controller. It will process data it receives from the GPS module.

GSM Module

SMS notifications will be sent to the user's emergency contacts using the GSM module. It will use the serial connection to link up to the Arduino board.

Emergency Button

To activate the safety alert, press the emergency button on the gadget. A digital input pin will be used to link the button to the Arduino board.

LED Indicators

To show the device's state, LED indicators will be used. The LED, for instance, blinks continually while the gadget is turned on and quickly when an emergency alert is sent out.

Buzzer

To notify the user that an emergency alarm has been activated, the device will include a buzzer.

Battery power will be used to run the device. The Arduino board and other components will be linked to the battery.

PROPOSED SYSTEM

Arduino Nano

The ATmega328P-based Arduino Nano is a small and adaptable microcontroller board. It is intended for small-scale projects that need a board with many connecting choices that are low-profile. The board has a 16 MHz quartz crystal oscillator, 8 analog input pins, and 14 digital input/output pins. Additionally, it has a DC power jack that can take a 7–12V input as well as a USB interface forprogramming and power. The board can beprogrammed using the Arduino IDE and is compatible with the majority of Arduino shields. Its compact size and adaptability make it a popular option for do-it-yourself projects, including GPS-enabled safety gadgets for women.

GSM 800C

The capability of GSM 800C to offer voice and data services to mobile devices is one of its primary features. To make optimum use of the available frequency spectrum, it combines time-division multiple access (TDMA) and frequency-division multiple access (FDMA) approaches. GSM 800C can carry data at speeds of up to 9.6 kbps, which is adequate for email and routine web browsing. Additionally, it enables SMS (short message service), which has gained popularity as a global form of communication. The security of GSM 800C is another crucial characteristic. Calls and data transmissions are safeguarded from unauthorized access using a range of encryption and authentication protocols.

Neo6M GPS Module

A well-liked GPS component with many uses is the NEO-6M. The Global Positioning System (GPS) satellites can be used by this small, low-power device to produce precise positioning and timing data. The module offers a high update rate of up to 5Hz and can accommodate up to 50 channels. Additionally, it has a built-in backup battery, allowing for a quicker time to first repair and dependable performance even in difficult circumstances. The NEO-6M module may produce data in both common and unique forms and employs the NMEA protocol for communication.

Rf Transmitter

A radio frequency (RF) transmitter is a piece of electronic equipment that creates radio waves and transmits them *via* an antenna into the atmosphere. An oscillator, a modulator, and an amplifier are among the common parts found in transmitters. The modulator modifies the signal to carry information such as voice, music, or data while the oscillator creates a carrier signal at a certain frequency. The signal is amplified by the amplifier until it is strong enough to be sent through the antenna. An RF transmitter's specs typically include the frequency band it operates in, the output power and the modulation technique it employs. The transmitter's access to the radio spectrum is determined by its frequency range, and the signal's range is determined by its output power.

Rf Receiver

An electronic device used to receive and process radio frequency signals is known as a radio frequency (RF) receiver. To extract information from modulated RF waves, it is frequently employed in wireless communication systems. The performance of an RF receiver in terms of its capacity to receive and process RF signals is determined by its specifications, which include frequency range, sensitivity, selectivity, noise figure, dynamic range, and linearity.

Other Components

- Buzzer
- Jumper wires
- Breadboard
- Female to male headers
- Male-to-male headers
- 9V battery

Fig. (**4**) shows a breadboard with wires and electronic components and Fig. (**5**) shows a proposed system's components mounted on the PCB. Setting up the development environment for the Arduino Nano involves installing the Arduino IDE and configuring the necessary parameters. This initial step is crucial for the subsequent stages of creating and implementing code to manage the device's functionalities. Upon completing the IDE setup, the development process focuses on crafting the code to handle various capabilities. These capabilities encompass tasks such as receiving GPS data, monitoring the panic button, and orchestrating responses like triggering alarms or buzzers. The code is pivotal in ensuring seamless integration and efficient functioning of the safety device. The system comprises two essential components: the transmitter and the receiver. The transmitter operates with an external power source, typically a 9V to 12V battery. On the other hand, the receiver is equipped with an SOS button. Integrating the panic button with the Arduino board enables users to swiftly access and activate it during times of distress. In the receiver component, a meticulous process is followed. The pin is initialized, setting the foundation for its subsequent functionalities. This pin is then connected to the digital pin of the Arduino Nano, establishing a crucial link for communication. Additionally, a third pin is employed to establish a connection with the breadboard's ground, ensuring proper grounding for the system. This well-organized approach to system setup and code development reflects the systematic and thoughtful design of the safety device. Each step contributes to the overall functionality, ensuring that the safety device, with its panic button and other features, operates seamlessly to provide swift assistance when needed.

Fig. (4). Proposed system's wiring done on the breadboard.

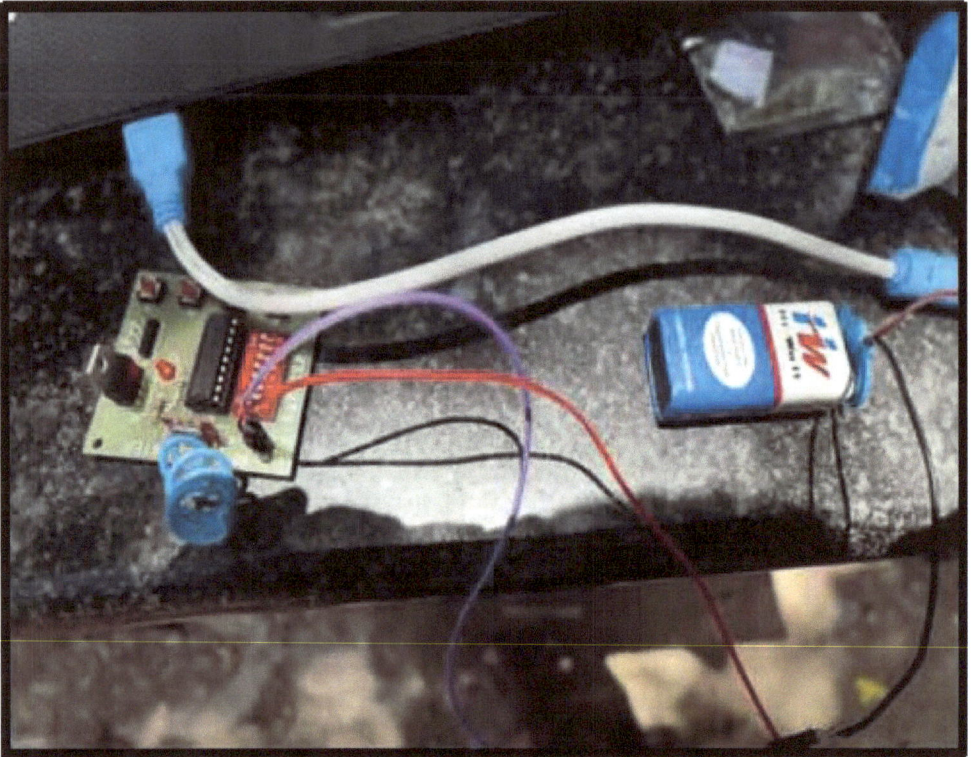

Fig. (5). Proposed system.

Setting up the development environment is the initial step in the installation process, involving the installation of the Arduino IDE. Once the environment is configured, the next phase involves coding to manage various functionalities of the system. This includes handling GPS data, monitoring the panic button, and triggering the alarm or buzzer when necessary. The system comprises two ends: a transmitter and a receiver. The transmitter relies on an external power source, typically a battery with a voltage ranging from 9V to 12V. On the other hand, the receiver incorporates an SOS button. The integration of the panic button with the Arduino allows users to quickly access and activate it in times of need. In the receiver component, the initialization of the pin is the first step. Subsequently, it is connected to the digital pin of the Arduino Nano, and a third pin establishes a connection to the ground on the breadboard. When interfacing a GPS module with an Arduino for acquiring location information, the TX and RX pins are initialized. These pins are then connected to the digital pins (4, 5) of the Arduino Nano, and the GPS module is powered using the VCC pin of the Arduino Nano. Similarly, when connecting a GSM module to an Arduino, the RX and TX pins are initialized. Following initialization, they are connected to the digital pins (2, 3) of the Arduino Nano. Power for the GSM module is provided externally, and power banks are commonly used for this purpose. This comprehensive setup ensures effective communication between the various modules, allowing for the seamless operation of the safety system. The utilization of power banks as an external power source enhances the system's portability and makes it suitable for a variety of applications, particularly in scenarios where quick access to safety features is paramount.

The integration of a GPS module with an Arduino Nano plays a pivotal role in providing accurate location information. This is achieved by initializing and connecting the TX and RX pins of the GPS module to the digital pins (4, 5) of the Arduino Nano. The GPS module is powered by the Arduino Nano's VCC. This connection allows the Arduino to receive location data from the GPS module, enabling precise tracking of the device's location. Similarly, the GSM module is connected to the Arduino Nano to facilitate communication. The RX and TX pins of the GSM module are initialized and then connected to the digital pins (2, 3) of the Arduino Nano. The external power source, such as a power bank, is employed to ensure continuous and reliable power to the GSM module. In the receiver component of the system, an SOS button is incorporated. The pin of the SOS button is initialized and connected to the digital pin of the Arduino Nano. This configuration enables users to easily access and activate the SOS button in times of distress. The alarm or buzzer, a critical component for signaling emergencies, is connected to the Arduino Nano's digital pin (7). In the event of an emergency, the buzzer is programmed to emit loud sounds, alerting those in the vicinity. The entire system operates in a synchronized manner. When the SOS button is pressed on the transmitter end, the transmitter sends out signals. These signals are then received by the RF receiver on the receiver end. The Arduino Nano on the receiver end, programmed to detect these signals, commands the GSM module (GSM 800c) to send SMS alerts to pre-registered phone numbers. Simultaneously, the GPS NEO6M module receives a command to track the location of the victim, providing real-time location data. Additionally, the buzzer is activated, emitting audible signals to alert nearby individuals. This comprehensive system ensures that in emergencies, not only are SMS alerts sent to designated contacts, but the location of the victim is also tracked, and audible alerts are sounded for immediate attention. The utilization of Arduino Nano and various modules showcases an effective and integrated approach to women's safety through advanced technology.

RESULTS & DISCUSSION

Fig. (**6**) shows the Simulation circuit and Fig. (**7**) shows the output of the simulation process. Fig. (**8**) shows the location of the person asking for help. The proposed safety device, designed to enhance women's security, offers versatile applications that extend beyond individual safety concerns. Its integration with existing safety infrastructure, such as police stations, emergency response teams, and hospitals, significantly enhances the potential for swift and effective emergency responses. By linking the device to established safety networks, including emergency services and law enforcement agencies, response times can be notably expedited, thereby strengthening the overall safety ecosystem. Furthermore, the device possesses the capability to establish community safety networks. By connecting multiple devices to a central server managed by a security team, a community-based safety network can be established. In times of crisis, this setup enables the security team to promptly respond by dispatching assistance to the location where the device has been activated. This collaborative approach not only extends the impact of the safety device beyond individual users but also cultivates a heightened sense of community safety. For remote workers, such as field workers, delivery drivers, and individuals in isolated environments, the safety device provides an additional layer of security. By configuring the device to detect deviations from predetermined routes or activities, it serves as a safeguard for those operating in remote or unfamiliar settings. This real-time monitoring and assistance capability contributes to enhancing the safety and well-being of individuals in potentially high-risk situations. Moreover, the compatibility of the safety device with wearable technology, such as fitness bands and smartwatches, allows for continuous safety monitoring. This integration facilitates seamless incorporation into individuals' daily routines, ensuring ongoing vigilance over their safety without disrupting their activities. In summary, the proposed safety device not only addresses individual security concerns but also integrates effectively with broader safety infrastructures and community networks. Its application extends to remote work environments, offering real-time monitoring and assistance, and integrates seamlessly with wearable technologies for continuous safety oversight.

FUTURE SCOPE

The safety device can be enhanced by integrating additional sensors. Sensors measuring parameters like temperature, humidity, and air quality can be connected to the system. This integration enables the device to identify potentially dangerous situations, such as extreme weather conditions or poor air quality, and promptly alert the wearer. By expanding the range of sensed parameters, the device becomes a comprehensive safety tool, addressing various environmental risks.

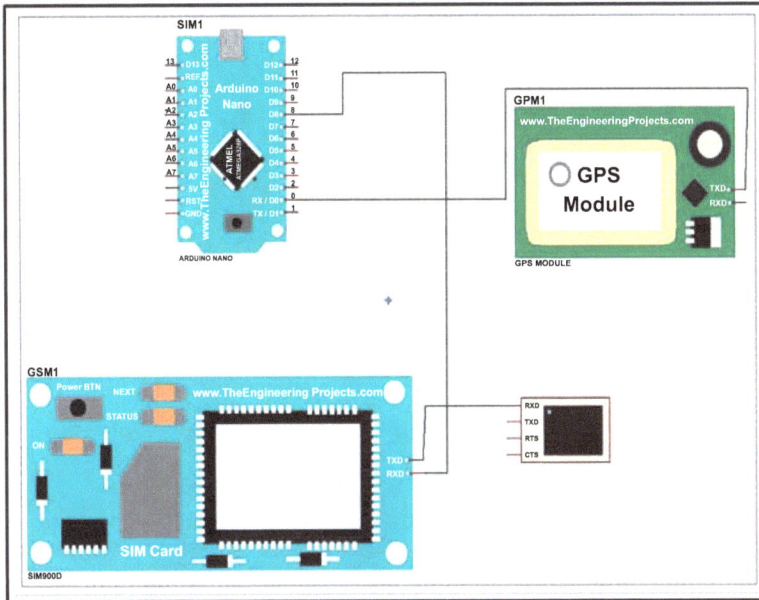

Fig. (6). Simulation circuit.

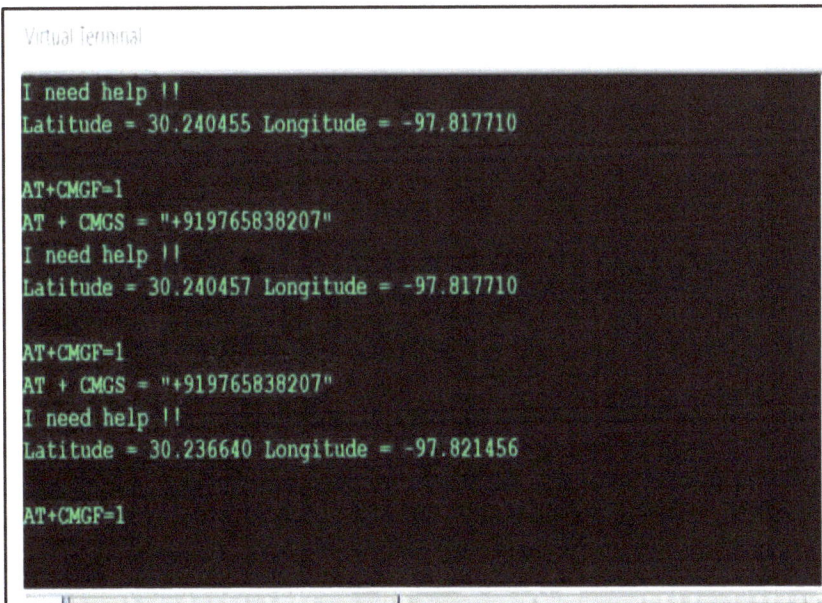

Fig. (7). Output of the simulation process.

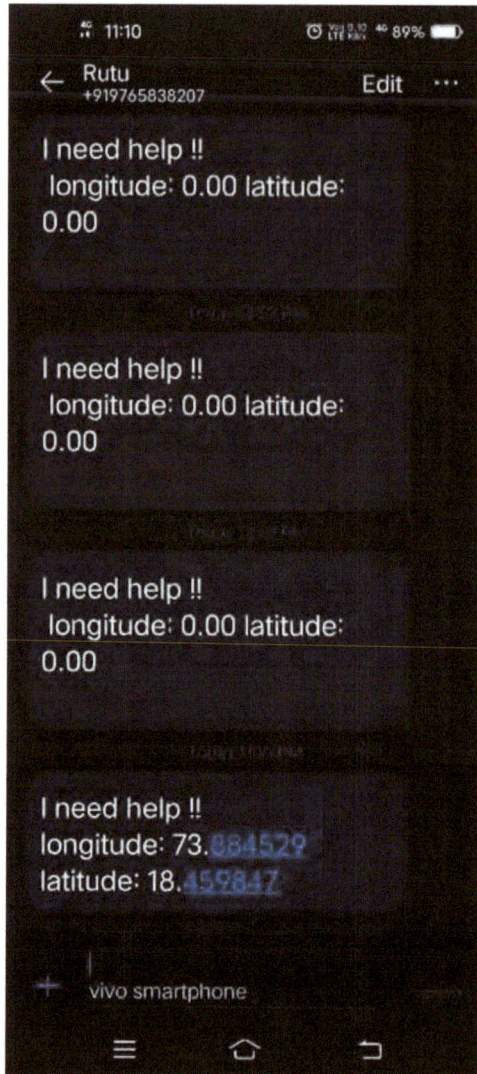

Fig. (8). Message showing the location of the person asking for help.

CONCLUSION

The integration of Arduino and GPS tracking technology in creating a women's safety device opens up numerous possibilities for addressing safety concerns and providing women with a heightened sense of security. This innovative device is designed to enhance human safety by offering real-time position monitoring and distress signal activation. By incorporating Arduino technology, a versatile and widely used microcontroller platform, this safety device becomes a powerful tool for ensuring the well-being of women in potentially hazardous situations. The

Arduino Nano, equipped with an ATmega328P microcontroller, serves as the central processing unit for the device, enabling seamless integration of various sensors and communication modules. The GPS tracking system embedded in the device plays a crucial role in ensuring accurate and real-time location monitoring. The GPS module, such as the NEO6M, communicates with the Arduino Nano to relay precise location data. This feature is fundamental in emergency situations, allowing for swift response and assistance. The SOS button, another integral component of the device, provides a quick and accessible means for users to activate distress signals. The emergency button is connected to the Arduino Nano, and when pressed, it initiates a chain of actions that include sending SMS notifications to pre-registered phone numbers and triggering audible alerts through a buzzer. Moreover, the device's compatibility with wearable technology, such as fitness bands and smartwatches, expands its usability and reach. This integration allows for continuous monitoring of the wearer's safety, offering a proactive approach to personal security. The device's potential for integration with existing safety infrastructure, including police stations, emergency response teams, and hospitals, is a significant advancement. This connectivity ensures that emergency responses can be expedited, contributing to faster and more efficient crisis management. The creation of a community safety network is another innovative aspect of this device. By connecting multiple devices to a central server controlled by a security team, a network is established for prompt response to crises. This community-oriented approach enhances overall safety within a locality. Furthermore, the device can be employed to ensure the safety of remote workers, such as field workers and delivery drivers, by recognizing deviations from their intended paths. This application extends the utility of the safety device beyond personal use to cover a broader spectrum of occupational safety concerns. The potential for integrating additional sensors, and measuring parameters like temperature, humidity, and air quality, enhances the device's capabilities. This multi-sensor integration allows the device to identify and alert users to potentially dangerous situations, enabling them to take proactive measures. In conclusion, the use of Arduino and GPS tracking technology in the creation of a women's safety device signifies a significant step towards addressing safety challenges and fostering a sense of security. The device's features, ranging from real-time location monitoring to integration with wearable technology and community safety networks, showcase its potential to positively impact women's safety in diverse scenarios.

REFERENCES

[1]   Kar, Rajib Kumar, A. S. S. Murugan, D. Obulesu, and Ch Shravani. "Design and practical application of a cost effective intelligent female surveillance system using GPS, GSM, and arduino technology." In *E3S Web of Conferences*, Vol. 472, p. 07, 2024.

[2]   Farooq, Muhammad Shoaib, Ayesha Masooma, Uzma Omer, Rabia Tehseen, S. A. M. Gilani, and

Zabihullah Atal. "The role of IoT in woman's safety: A systematic literature review." *IEEE Access*, Vol. 11, pp. 69807-69825, 2023.

[3]   B. Vamshikrishna Yadav, A. Viji Amutha Mary, M. Paul Selvan, S. Jancy and L. S. Helen, "Arduino based Women Safety Tracker Device," *7th International Conference on Trends in Electronics and Informatics (ICOEI)*, Tirunelveli, India, pp. 433-436, 2023. [http://dx.doi.org/10.1109/ICOEI56765.2023.10126053]

[4]   Bakka, Praveen, Rishi Mullagiri, Yashwanth Reddy Purra, Sri Sai Bhargav Vudatha, Anil Kumar Budati, and Lakshmi Prasad Kolalapudi. "Women security device using cellular networks with message alerting system." *AIP Conference Proceedings*, vol. 2794, no. 1, 2023. [http://dx.doi.org/10.1063/5.0165847]

[5]   Ananthula, Navya, T. Rajeshwari, B. Mounika, PA Harsha Vardhini, and B. Kalyani. "Arduino-based Rescue device with GPS Alert for Women Safety Application." *International Mobile and Embedded Technology Conference (MECON)*, pp. 343-347. 2022.

[6]   Swapnarani, P., P. Ramchandar Rao, and Vinit Kumar Gunjan. "Self-defense system for women safety with location tracking and SMS alerting through GPS and GSM networks." *Modern Approaches in Machine Learning & Cognitive Science: A Walkthrough*, Cham: Springer International Publishing, pp. 361-368, 2022.

[7]   Saranya, N., R. Aakash, K. Aakash, and K. Marimuthu. "A Smart friendly IoT device for women's safety with GSM and GPS location tracking." *5th International Conference on Electronics, Communication and Aerospace Technology (ICECA)*, pp. 409-414., 2021.

[8]   N. Prakash, E. Udayakumar, N. Kumareshan, and R. Gowrishankar, "GSM-based design and implementation of women safety device using Internet of Things", *Proceedings of ICBDCC 2019*, pp. 169-176, 2021. [http://dx.doi.org/10.1007/978-981-15-5285-4_16]

[9]   Kalyani, T. V. Sai, V. Mounika, P. Pooja, V. Sai Sahith, B. Pranay Kumar, and C. Akhil Kumar. "A novel approach to provide protection for women by using smart security devices." *Int. Res. J. Eng. Technol. (IRJET)*, vol. 7, no. 05, 2020.

[10]   A. Ranganadh, "Women safety device with GPS tracking and alerts", *Proceedings of the 4th ICIEEE*, pp. 797-805, 2020. [http://dx.doi.org/10.1007/978-981-15-2256-7_74]

[11]   G. Gulati, B.P. Lohani, and P.K. Kushwaha, A novel application of IoT in empowering women safety using GPS tracking module.*Res. Innov. Knowl. Manag. Technol. Appl. Bus. Sustain. (INBUSH)*, pp. 131-137, 2020. [http://dx.doi.org/10.1109/INBUSH46973.2020.9392193]

[12]   Sunehra, Dhiraj, V. Sai Sreshta, V. Shashank, and B. Uday Kumar Goud. "Raspberry Pi based smart wearable device for women's safety using GPS and GSM technology." *International Conference for Innovation in Technology (INOCON)*, p. 5, 2020.

[13]   R. Ramachandiran, L. Dhanya, and M. Shalini, "A survey on women safety device using IoT", *International Conference on System, Computation, Automation and Networking (ICSCAN)*, p. 6, 2019. [http://dx.doi.org/10.1109/ICSCAN.2019.8878817]

[14]   K. Seelam, and K. Prasanti, "A novel approach to protect women by using smart security device", *2nd International Conference on Inventive Systems and Control (ICISC)*, pp. 351-357, 2018.

[15]   Bhavale, Ms Deepali M., Ms Priyanka S. Bhawale, Ms Tejal Sasane, and Mr Atul S. Bhawale. "IoT based unified approach for women and children security using wireless and GPS." *Int. J. Sci. Dev. Res. (IJSDR)*, Vol. 3, no. 10, 2018.

CHAPTER 10

Ethical Data Practices in Digital Health: A Patient-Centric Future in Society 5.0

Rakesh Nayak[1,*], **Umashankar Ghugar**[2] and **P. Sudam Sekhar**[3]

[1] *Department of CSE, OP Jindal University, Raigarh, India*

[2] *Department of CSE, University Institute of Engineering (UIE), Chandigarh University, Mohali, Punjab, India*

[3] *Department of Mathematics and Statistics, Vignan University, Guntur, India*

Abstract: The healthcare sector has seen a change because of the quick development of digital health technologies, which provide a wealth of advantages and prospects. But with all of this digital change, there are also significant ethical questions about how healthcare data is gathered, stored, and used. This chapter examines the relationship between data ethics and digital health, emphasizing the main obstacles to ethical and responsible activities as well as possible solutions. It explores issues including informed consent, patient privacy, data security, and the moral ramifications of AI and machine learning algorithms [1] in the healthcare industry. This chapter intends to provide a greater knowledge of the significance of data ethics in creating a reliable and patient-centric digital healthcare ecosystem by exploring the ethical aspects of digital health.

Keywords: Consent and transparency, Data privacy, Data security, Digital health, Ethical consideration.

INTRODUCTION

Over the last few years, the electronic change in the medical care sector has brought about a substantial rise in the collection, and storage space coupled with the use of individual information. While this change has produced various advantages it has likewise increased worries concerning the ethical consequences of managing delicate wellness details. As the assimilation of modern technology in medical care continues to progress, the value of ethical data practice in electronic health cannot be exaggerated.

* **Corresponding author Rakesh Nayak:** Department of CSE, OP Jindal University, Raigarh, India;
E-mail: nayakrakesh8@gmail.com

Parikshit N. Mahalle, Gitanjali R. Shinde, Namrata N. Wasatkar & Prashant R. Anerao (Eds.)

Most importantly ethical data practices are crucial for promoting individual privacy as well as discretion. In the electronic age, substantial quantities of individual wellness information are being created and shared throughout different systems consisting of digital wellness documents, wearable tools, as well as telemedicine applications. It is important that medical care companies along with modern technology suppliers comply with stringent moral criteria to make certain that individual information is protected from unapproved accessibility, abuse, or exploitation.

Furthermore, ethical information methods play a vital function in advertising openness as well as responsibility within the electronic health landscape. Medical care suppliers as well as modern technology designers need to be clear regarding exactly how individual information is gathered, saved as well as made use of, in addition to the objectives for which it is being used. Moreover, clear and easily accessible plans pertaining to information administration and grants equip individuals to make enlightened choices regarding the sharing of their wellness details. By developing durable ethical structures and administration devices, the liable handling of information can be made sure therefore mitigating the threat of information violations as well as unapproved information handling.

In addition, the ethical use of information in electronic wellness is carefully connected to the innovation of clinical research study and the distribution of tailored treatment. By leveraging de-identified and accumulated wellness information in a liable fashion, scientists along with healthcare experts can obtain useful understandings right into condition patterns, therapy results as well as populace wellness patterns. Honest information techniques make it possible for the accountable sharing of information for research study objectives adding to the growth of evidence-based medical treatments as well as the renovation of individual results. Additionally, the ethical evaluation of specific individual information can sustain the shipment of tailored as well as accurate medication promoting customized therapy that straightens with the special demands of people.

The ethical information is basic to the ethical and also accountable improvement of electronic wellness. By maintaining concepts of personal privacy, openness as well as responsibility, healthcare stakeholders can grow an atmosphere of dependence along with honesty in making use of client information. Accepting ethical information methods not only safeguards clients' privacy but also fosters advancementand distribution of patient-centered treatment [2]. As electronic wellness continues to transform the medical care landscape, the honest factors to consider bordering information techniques have to continue to be at the center of sector campaigns making certain the ethical structure of electronic wellness for the advantage of clients and culture as a whole.

CONTRIBUTION OF THE CHAPTER

• To explore the different patient-centric approaches to data in digital health
• To know different ethical data practices in digital health
• To discuss the ethical considerations in the future of digital health
• To discuss best practices for data handling in digital health

This chapter has been organized into nine sections. The first section describes the motivation for this chapter. The second section represents the patient-centric approach to handling data in digital health care. The third section represents ethical practices for ensuring privacy and trust. Section third represents the ethical data practices in the digital health landscape, and the fourth section describes ethical considerations in the future of digital health. The fifth and sixth section describe ethical data practices and empowering patients in digital health care. The seventh section represents the role of ethical data practices. Best practices for data handling are described in section eighth. Balancing data practices and privacy is described in section nine followed by the conclusion.

MOTIVATION

We envision a future in which technical developments in electronic health and wellness have permitted it to be efficiently incorporated right into every component of our lives. In this instance, individuals have access to personalized wellness information allowing them to make educated choices concerning their health and wellness.

The motivation of this article comes to be noticeable in this setup. As the situation plays out, we witness a culture in which individuals' health and wellness details are valued not simply for their wellness but likewise for the enhancement of public wellness projects coupled with clinical study.

The circumstance establishes the phase for discovering the vital significance of personal privacy, permission, as well as openness in the collection, storage space, and utilization of health and wellness information. It highlights the possible advantages of leveraging this information for customized medication and also populace health and wellness monitoring while highlighting the demand for durable moral structures to guarantee that people's civil liberties and freedom are shielded.

In this scenario, the section intends to motivate visitors to proactively participate in forming the moral landscape of electronic health and wellness in Culture 5.0. It motivates stakeholders throughout health care, innovation, and policy-making as well as the public to acknowledge their functions in promoting a patient-centric future where moral information methods drive favorable results for people as well as culture as a whole.

PUTTING PATIENTS FIRST: A PATIENT-CENTRIC APPROACH IN DIGITAL HEALTH

In the quickly advancing landscape of electronic wellness, the emphasis is progressively changing in the direction of an individual-centric strategy for information. This standard change holds the possibility to change health care, by encouraging clients enhancing end results, and driving technology. Right here is an extensive take a look at the value of placing people initially in the world of electronic wellness. Fig. (**1**) illustrates a patient-centric approach to digital health, emphasizing user engagement and personalized care.

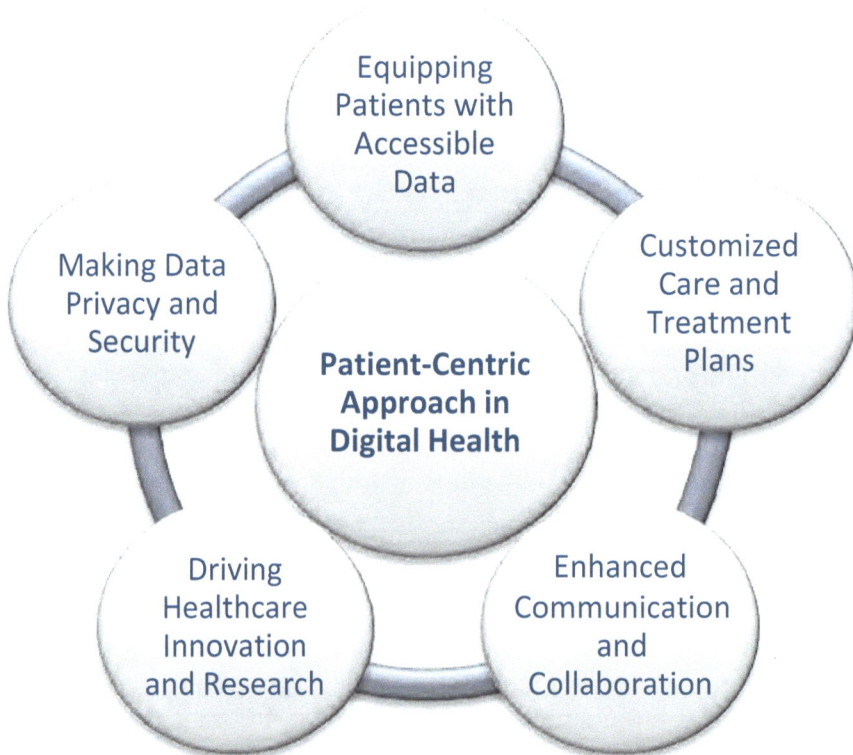

Fig. (1). A patient-centric approach in digital health.

- **Equipping Patients with Accessible Data:** Supplying clients with access to their wellness information [3] cultivates openness and equips them to take an energetic function in their treatment. Obtainable information allows individuals to make educated choices, causing far better health and wellness monitoring and improved interaction with doctors.
- **Customized Care and Treatment Plans [4]:** Leveraging personal information enables the growth of tailored treatment strategies customized to the person's special needs as well as choices. Individualized therapy intended based upon personal information can bring about a lot more reliable treatments, boosted conformity along eventually much better wellness results.
- **Enhanced Communication and Collaboration:** A patient-centric strategy to information cultivates enhanced interaction and partnership between individuals as well as doctors [5]. Obtainable information helps with purposeful conversations and educated decision-making as well as shared choice procedures, reinforcing the patient-provider connection.
- **Driving Healthcare Innovation and Research:** Patient-generated information in electronic health and wellness development drives innovations in therapy methods along with healthcare innovations [6].
- **Making Data Privacy and Security [7]:** Prioritizing patient-centric information includes extensive actions to secure personal privacy as well as protection, ensuring the privacy and stability of individual wellness information. Robust data protection mechanisms are essential to building and maintaining patient trust in digital health solutions.

Welcoming a patient-centric method to information in electronic wellness is crucial for cultivating individual empowerment, tailored treatment, boosted interaction, health care advancement as well as making sure the personal privacy as well as safety of person info. By putting people at the core of the electronic wellness ecological community we can lead the way for an extra efficient, reliable as well as caring medical care system.

ENSURING PRIVACY AND TRUST: ETHICAL DATA PRACTICES IN DIGITAL HEALTH LANDSCAPE

The development of electronic wellness innovations has resulted in a transformative change in the healthcare sector, using unmatched chances to improve personal treatment, enhance end results, as well as improve healthcare shipment. Nonetheless, as the industry continues to accept electronic development it is important to focus on honest information techniques to guard individual personal privacy, promote count on, as well as promote the highest degree of information administration.

Within the vibrant landscape of electronic health, the honest collection, usage, and security of individual information are vital to cultivating depending on making sure personal privacy. This write-up looks into the relevance of honest information techniques in the electronic health landscape highlighting the critical function they play in protecting client personal privacy and developing count. Fig. (**2**) showcases ethical data practices in the digital health landscape, promoting privacy and responsible data handling.

Fig. (2). Ethical data practices in digital health landscape.

- **Valuing Patient Privacy in the Digital Age:** The electronic wellness landscape offers distinct difficulties coupled with possibilities in regard to individual personal privacy [7]. Honest information techniques are vital in making sure that client personal privacy is maintained in the collection, and storage space as well as the use of health and wellness information hence fostering complacency and dependence amongst individuals.
- **Information Governance and Compliance:** Developing durable information administration structures and also making sure conformity with information

defense laws are important parts of ethical information methods in electronic health and wellness. Sticking to rigorous information governance requirements not only safeguards a person's privacy but also shows a dedication to honest conduct within the electronic health and wellness environment.

- **Information Consent and Transparency [8]:** Valuing personal freedom *via* educated permission and openness pertaining to information collection and use is essential to ethical information methods. People need to have a clear understanding of exactly how their information will certainly be used and their authorization must be acquired in a way that is relatively easy to fix.

- **Minimizing Data Breach Risks:** Proactively attending to information protection threats and executing durable actions to reduce the capacity for information violations belong to promoting ethical data practice. By focusing on information safety and electronic wellness, stakeholders can instill self-confidence in individuals as well as decrease the danger of unapproved accessibility to delicate wellness information [9].

- **Equipping Patients with Data Ownership:** Ethical information techniques include encouraging clients to work out possession as well as control over their health and wellness information. Offering clients the capability to access, handle as well and share their wellness details improves openness and fosters a feeling of freedom, therefore enhancing the patient-provider partnership [10].

Ethical data practices form the cornerstone of preserving patient privacy and fostering trust within the digital health landscape. By focusing on information administration, approval, safety, and security procedures, coupled with client permission, stakeholders can browse ethical intricacies of information in electronic wellness inevitably adding to a medical care community improved personal privacyas well as regard for individual well-being.

ETHICAL CONSIDERATIONS IN THE FUTURE OF DIGITAL HEALTH

As we depend on the brink of a brand-new period in medical care, the power of information in changing the electronic health and wellness landscape is undeniable. The merging of modern technology and healthcare has caused a wide range of information to be produced providing extraordinary chances to enhance client treatment, improve clinical research study, as well as drive technology. Nonetheless, in the middle of this information transformation, honest factors to consider are vital in making certain that the prospective advantages are optimized while protecting the client's personal privacy count on.

The quick development of electronic wellness innovations has opened extraordinary capacity in leveraging information to reinvent healthcare distribution. As information becomes increasingly more important to the future of medical care, it is necessary to promote ethical factors. Fig. (**3**) illustrates ethical guidelines outlining principles and rules for safeguarding sensitive information.

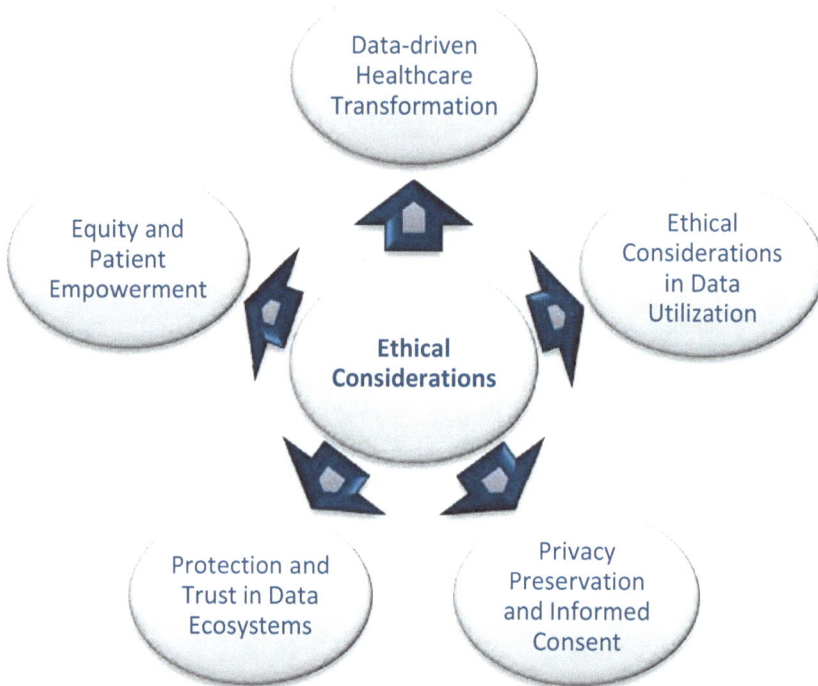

Fig. (3). Ethical consideration for digital health landscape.

- **Data-driven Healthcare Transformation:** The merging of innovative analytics, expert systems, and also adjoined wellness systems has brought about a period of data-driven healthcare improvement. From tailored therapy routines to anticipating analytics for very early discovery, the capacity of information to boost individual end results is enormous. Nonetheless, this change demands a conscientious technique to ethical data practice to alleviate possible dangers and a secure client rate of interest [11].

- **Ethical Considerations in Data Utilization:** The ethical utilization of health data encompasses multifaceted considerations, including privacy protection, consent management, data security, and equitable access. Respect for patient autonomy and privacy must guide the collection, storage, and sharing of health data, ensuring that individuals retain control over their personal health information. Moreover, transparency in data usage and the equitable distribution of benefits derived from data-driven insights are pivotal ethical imperatives [12].

- **Privacy Preservation and Informed Consent:** Protecting a person's privacy in the electronic wellness landscape needs durable information administration structures, security procedures, and stringent accessibility controls. Honest information methods require the application of devices that promote client confidentiality while making it possible for information access for genuine healthcare functions. Informed permission, defined by openness and understanding encourages people to make independent choices pertaining to using their wellness information, consequently enhancing ethical information application [13].

- **Protection and Trust in Data Ecosystems:** The ethical importance of information safety cannot be overemphasized in the context of electronic wellness. Minimizing the dangers of information violations, unapproved access as well as cyber risks is necessary to encourage self-confidence in information ecological communities. Stakeholders have to focus on the application of resilient safety and security procedures, information file encryption, and adherence to regulative requirements to maintain the honesty and integrity of wellness information [14].
- **Equity and Patient Empowerment:** Ethical data practices require a dedication to minimizing variations in information accessibility making certain that underserved populaces profit equitably from data-driven health care advancements. In addition, encouraging clients with accessibility to their health and wellness information in addition to cultivating a participatory function in decision-making procedures supports the honest structures of information usage, provoking an individual-centric technique within the electronic wellness landscape.

The future of electronic health and wellness holds an enormous pledge to leveraging information to drive transformative innovations in medical care. Nevertheless, this needs to be underpinned by unwavering honest factors to consider a person's well-being, personal privacy preservation, and a trust fund. By welcoming honest structures that promote the concepts of beneficence, freedom, and also justice, stakeholders can browse the intricacies of information usage in electronic wellness, promoting an environment that takes advantage of the power of information sensibly and fairly for the betterment of worldwide medical care.

ETHICAL DATA PRACTICES FOR DIGITAL HEALTH INNOVATION

In the ever-evolving landscape of electronic wellness, ethical information methods are necessary for cultivating development while protecting individual personal privacy. As innovation continues to change the healthcare market, the responsible use of information is extremely important to guarantee that electronic wellness advancement continues to be lasting, impactful, and ethical.

The crossroads of health care and innovation have generated a wide range of health-related information, consisting of digital wellness documents, wearable tools, telemedicine systems, and wellness applications. This wealth of information provides extraordinary chances for enhancing patient treatment, progressing clinical research studies, as well as driving advancement. Nonetheless, with these possibilities come ethical factors to consider that need to be thoroughly explored to make sure that the advantages of electronic wellness development are optimized without endangering individual personal privacy and protection.

At the heart of ethical information methods in electronic wellness advancement is the critical focus on personal privacy as well as information safety and security. As medical care companies and modern technology businesses accumulate, shop, as well as evaluate large quantities of wellness information, it is essential to apply durable information security procedures. This consists of rigorous file encryption procedures, accessibility controls, and adherence to personal privacy laws to stop unapproved gain access as well as information violations. By focusing on the safety and security of personal information, electronic wellness pioneers can construct a structure of trust coupled with self-confidence amongst individuals and medical care stakeholders.

Openness is an additional basic aspect of ethical information techniques in electronic health and wellness technology. Individuals need to be notified concerning just how their information will certainly be made use of, who will certainly have access to it, and what procedures remain in place to safeguard their privacy. Clear and thorough grant devices are crucial in encouraging clients to make educated choices concerning making use of their wellness information. By advertising openness, electronic wellness pioneers can grow a society of responsibility as well as trust, promoting collective connections with individuals as well as medical care service providers.

Liable information usage for study and technology is a crucial ethical factor to consider in the electronic wellness landscape. While accumulated and de-identified information can drive innovative clinical explorations and populace health and wellness understandings, the procedure of de-identification needs to be extensive to avoid re-identification and safeguard private personal privacy. Honest standards must be developed to control the use of personal information for the study, guaranteeing that it is used for the advantage of culture while appreciating specific freedom. By sticking to accountable information usage methods, electronic wellness trendsetters can utilize the possibility of information to drive significant developments in medical care while maintaining honest criteria.

A lasting future for electronic wellness technology likewise depends on a patient-centric approach to information. Encouraging individuals to have control over their wellness information, allowing them to gain access to, share, and handle their details, is essential to ethical information techniques. Individual sites, interoperable wellness systems, as well as data-sharing systems can help with client interaction with their wellness information, changing them into energetic individuals in their treatment. By positioning clients at the facility of information administration, electronic health, and wellness trendsetters can support the concepts of freedom and encourage people to take an energetic role in their health and wellness.

Constructing a lasting future for electronic wellness development calls for a steadfast dedication to ethical information techniques. By focusing on personal privacy, openness, liable information usage, and personal empowerment, electronic wellness trendsetters can drive purposeful development while supporting the trust fund and also health of individuals. It is with ethical data techniques that the possibility of electronic health and wellness advancement can be understood in a lasting and impactful way, making sure that technology and ethical factors to consider go together in the direction of a brighter as well as a much healthier future.

EMPOWERING PATIENTS IN THE DIGITAL HEALTH REVOLUTION

The electronic health change has catalyzed a standard change in medical care encouraging individuals with extraordinary accessibility to customized treatments along with energetic involvement in their health trips. As technical developments remain to redefine the healthcare landscape, the empowerment of individuals stands as a keystone of this transformative period. This post checks out the critical function of personal empowerment in the electronic wellness change, highlighting the diverse ways in which innovation is cultivating a patient-centric standard in healthcare distribution. Fig. (**4**) illustrates the main features of empowering patients in digital health through awareness, understanding, sharing, and wellness advocacy.

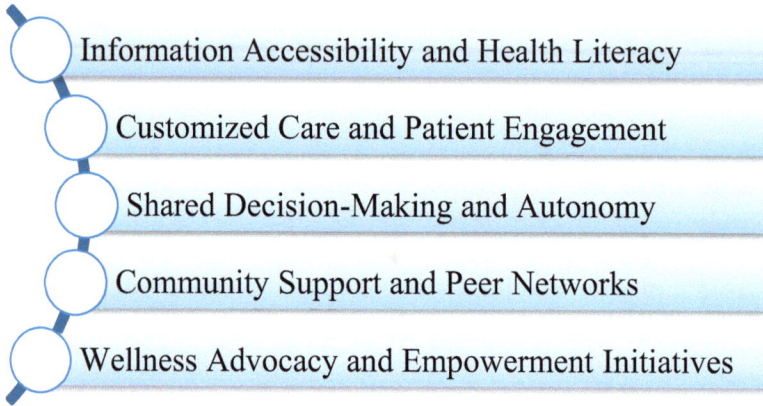

Fig. (4). Empowering patients in the digital health.

- **Information Accessibility and Health Literacy [15]:** The expansion of electronic wellness systems coupled with mobile applications has democratized accessibility to wellness details, making it possible for individuals to enlighten themselves concerning clinical problems, therapy alternatives as well as preventive treatments. By cultivating wellness proficiency and offering thorough sources of electronic health and wellness, modern technologies equip people to make educated choices concerning their wellness.
- **Customized Care and Patient Engagement:** Digital wellness services such as telemedicine [16] wearable tools, in addition to remote tracking devices have redefined the patient-provider connection helping with continual interaction and individualized treatment distribution. *Via* remote appointments real-time health and wellness information monitoring, and online treatment synchronization, individuals are encouraged to proactively take part in their treatment strategies, promoting a feeling of firmness and partnership in the administration of their health and wellness problems.
- **Shared Decision-Making and Autonomy [17]:** The electronic wellness transformation highlights the principles of shared decision-making, in which clients are urged to proactively take part in therapy conversations, express their choices, and deal with preparation and healthcare providers. This collective method not only recognizes client freedom but additionally promotes a feeling of empowerment making it possible for people to play an energetic role in forming their healthcare experiences and results.
- **Community Support and Peer Networks:** Social connection [18] and peer assistance play a critical role in client empowerment within the electronic wellness landscape. On the internet, assistance teams, and social media provide opportunities for people to share experiences, look for suggestions, and amass

psychological assistance promoting a feeling of uniformity coupled with empowerment among health and wellness obstacles. These online areas function as drivers for empowerment, supporting an encouraging community that goes beyond geographical borders.

- **Wellness Advocacy and Empowerment Initiatives:** Digital wellness systems have become drivers for wellness campaigning coupled with empowerment efforts making it possible for people to promote their demands, increase recognition concerning certain health and wellness concerns, and also drive plan modifications. *Via* social media site projects, online applications coupled with online campaigning for initiatives, individuals are leveraging electronic systems to magnify their voices, support medical care reform [19].

The electronic health and wellness change has introduced a brand-new period of personal empowerment, redefining standard medical care characteristics and putting people at the center of their treatment experiences. By utilizing the possibility of innovation to cultivate details, ease of access customized treatment, shared decision-making and area assistance individuals are presuming progressively energetic functions in their health trips. As electronic health and wellness continue to progress, the empowerment of individuals will certainly continue to be an essential concept driving fair accessibility to care, cultivating freedom, as well as generating a patient-centric standard that goes beyond standard healthcare limits.

THE ROLE OF ETHICAL DATA PRACTICES IN DIGITAL HEALTH

In the age of electronic wellness, the merging of innovation and health care has released a wave of development that assures a change in personal treatment as well as professional results. Central to this change is the moral use of client information, which offers a column for advertising openness and responsibility in electronic health campaigns. As the landscape of health care continues to develop, the honest factors to consider bordering information methods play a critical role in forming a future where client health, as well as personal privacy, issafeguarded. This section looks into the important function of ethical information techniques in electronic wellness and their effect on advertising openness and responsibility in the healthcare sector. Fig. (**5**) illustrates the role of ethical data.

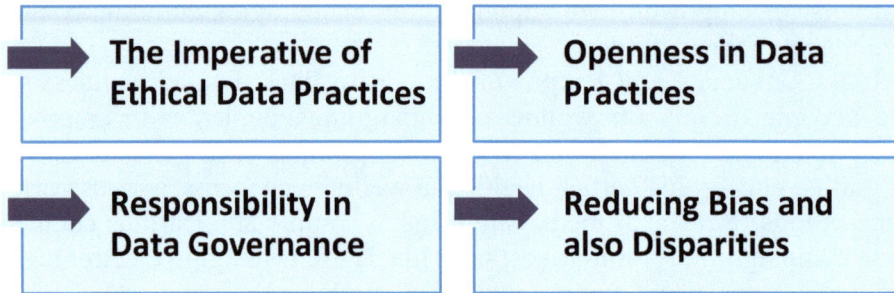

Fig. (5). Role of ethical data practices in digital health.

- **The Imperative of Ethical Data Practices:** The collection, storage space, as well as the use of personal information in electronic wellness, bring extensive ramifications for individuals'privacy, permission, safety and security. Thus ethical information methods are basic in promoting the trust fund along with the self-confidence of clients in the electronic wellness ecological community. By focusing on ethical data methods, healthcare stakeholders can show a dedication to appreciating personal freedom, securing delicate health and wellness details, as well as making certain that data-driven campaigns are carried out with the highest possible honest criteria [20].
- **Openness in Data Practices:** Transparency creates the foundation of ethical data practice in electronic health. Individuals can recognize just how their information is being used that has accessibility to it as well as the procedures in position to secure its privacy. Clear information methods equip clients to make educated choices regarding their engagement in electronic health efforts. Additionally, openness works as a driver for liability, engaging healthcare companies and modern technology suppliers to run with honesty and visibility in their data-related ventures.
- **Responsibility in Data Governance:** Ethical information methods require a durable structure of responsibility to make sure that personal information is used in a liable as well as lawful fashion. Healthcare companies as well as modern technology firms need to think of liability for the moral use of individual information, consisting of conformity with laws, sector requirements as well as finest techniques. Liability not only secures individual rates of interest but also instills self-confidence in the more comprehensive area that data-driven medical care advancements are regulated by honest concepts and also a dedication to a person's health.

- **Reducing Bias and Disparities:** Ethical information methods in electronic wellness necessitate a collective initiative to determine and alleviate predispositions as well as variations that might emerge in information collection and evaluation. As electronic health and wellness innovations significantly rely upon formulas as well as artificial intelligence, there is a threat of continuing medical care differences or accidentally victimizing particular individual populations. By proactively resolving predispositions and differences, stakeholders can function in the direction of cultivating an extra fair and comprehensive healthcare landscape where data-driven understandings profit all people irrespective of their history or group account.

Ethical information is important in advertising openness as well as responsibility in electronic health. By promoting moral criteria promoting openness as well as thinking of obligation for making use of personal information, the healthcare sector can grow into a society of integrity in the electronic health ecological community. Additionally, proactively dealing with predispositions along with differences in ethical data techniques can add to the awareness of a healthcare landscape that is fair, comprehensive, and fixated on individual health. As electronic health and wellness continue to break through, the moral use of individual information stands as a sign of obligation as well as honest stewardship assisting the sector in the direction of a future where development and patient-centric treatment exist harmoniously.

BEST PRACTICES FOR DATA HANDLING IN DIGITAL HEALTH

In the rapidly advancing landscape of electronic wellness, the honest handling of information is of critical value. As technical developments remain to improve the healthcare sector, the accountable collection, storage space, and application of personal information have become crucial honest factors to consider. This post intends to discover the most effective techniques for information handling in electronic wellness, highlighting the honest imperatives that highlight the liable monitoring of delicate health and wellness information. Fig. (**6**) describes the best practices for data handling in digital health. A triangular graphic with text on four levels discussing patient privacy: "securing patient privacy and consent," "compliance with regulatory standards," "information security and encryption," and "ethics with accountability.

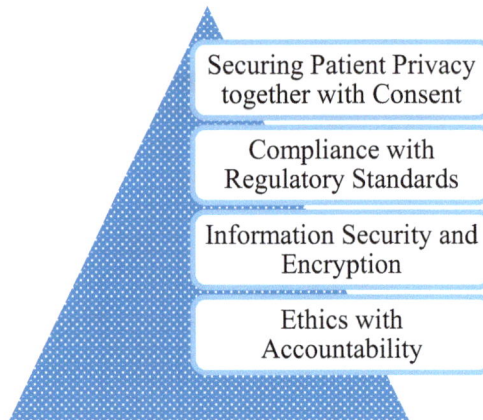

Fig. (6). Best practices for data handling in digital health.

- **Securing Patient Privacy and Consent:** At the core of ethical data handling in electronic wellness exists the security of individual personal privacy coupled with the regard for specific approval. Healthcare companies as well as innovation suppliers should focus on the application of durable personal privacy safeguards to make certain that personal information stays protected as well as private. Additionally getting educated authorization from people for the collection and use of their information is important to encourage people to make independent choices relating to the sharing of their wellness info.

- **Compliance with Regulatory Standards:** Compliance with regulative requirements and information defense regulations is basic to ethical data handling in electronic wellness. Healthcare stakeholders should browse a complicated internet of laws such as the Health Insurance Portability as well as Accountability Act (HIPAA) in the United States or the General Data Protection Regulation (GDPR) in the European Union, to make certain that information care techniques straighten with lawful demands. Adhering to these requirements not only supports moral concepts but likewise minimizes the threat of lawful as well as economic consequences originating from non-compliance [21].

- **Information Security and Encryption:** The honest handling of information in electronic wellness requires an undeviating dedication to information protection and security. Durable security steps need to be used to protect individual information from unapproved accessibility, violations, and cyber risks. By applying state-of-the-art safety methods, healthcare companies can show their commitment to promoting the honesty and privacy of individual details, thus promoting dependence among individuals along with stakeholders [22].

- **Ethics with Accountability:** Transparency, as well as responsibility, develops the bedrock of honest information dealing with methods. Health care companies as well as modern technology suppliers need to be clear concerning their information dealing with procedures consisting of information collection techniques, storage space techniques as well as information sharing methods. Furthermore, thinking responsibility for the moral administration of individual information is essential as it inspires self-confidence among clients and the more comprehensive area that their information is managed with stability and according to moral requirements [23].

Exploring the moral landscape of information handling in electronic health and wellness calls for an unfaltering dedication to the finest techniques that focus on client personal privacy governing conformity, information protection, openness, and honest usage for study and advancement. By sticking to these finest methods, medical care stakeholders can support moral criteria, and add to the development of electronic health and wellness in an accountable as well as moral fashion. As the electronic health and wellness landscape continues to progress, the moral handling of personal information continues to be an essential element of honest administration along with honesty in the search for boosted personal results together with healthcare shipment.

BALANCING DATA PRACTICES AND PRIVACY

In the swiftly advancing landscape of electronic health, the capacity for transformative development is enormous. From wearable tools that keep an eye on essential indicators to telemedicine systems that link clients with doctors, the combination of modern technology right into medical care has the power to enhance individual results as well as reinvent the distribution of clinical solutions. Nonetheless, as the electronic health change unfolds, the moral ramifications of information techniques are entering into sharp emphasis.

Among the main factors to consider in the electronic health landscape is the defense of a person's privacy. As medical care information becomes progressively digitized, there is an expanding requirement to protect delicate details from unapproved accessibility and abuse. Health care companies should execute durable information safety and security procedures such as security, access to controls and protected information storage space to secure individual privacy and stop violations that can endanger personal privacy.

In addition, the liable collection, storage space, and use of client information are vital moral imperatives in electronic wellness. Medical care companies as well as modern technology businesses need to follow rigorous standards for getting client authorization, transparently connecting information use techniques as well, and guaranteeing that information is utilized just for reputable medical functions. This consists of the honest handling of individual health and wellness details, hereditary information, and various other delicate information that can affect a person's privacy and health.

One more crucial facet of ethical data techniques in electronic wellness is the requirement for openness and responsibility. People need to be notified concerning exactly how their information will be made use of that will certainly have accessibility as well as the steps in position to shield their privacy. Additionally, medical care companies and innovation programmers have to be responsible for their information techniques, on a regular basis examining their systems as well as sticking to market requirements and laws to keep honesty.

In the quest for ethical data practice [24], it is crucial to strike an equilibrium between technology and personal privacy. While welcoming technical breakthroughs and data-driven understandings, it is important to maintain honest criteria that focus on individual well-being. This needs a positive strategy for moral decision-making, continual analysis of information techniques along with a dedication to maintaining the greatest honest criteria in the electronic health community.

As the electronic health landscape continues to develop, stakeholders should work together to develop and support honest standards that protect a person's privacy coupled with dependents. This consists of advertising a society of moral recognition buying durable information safety framework and focusing on person empowerment with educated authorization coupled with information openness.

In recap, the period of electronic wellness provides unparalleled possibilities for progression, however, it additionally requires a resolute dedication to honest information techniques. By focusing on client personal privacy, information safety, openness, and responsibility, the electronic health change can be utilized to its greatest possibility while promoting the moral requirements that underpin the count on the health of individuals. As we browse the moral landscape of electronic health, it is important to promote the equilibrium between development as well as personal privacy making certain that ethical data practice works as the keystone of a lasting as well as patient-centered electronic wellness future.

CONCLUSION

The fast growth of electronic health and wellness as well as health and wellness devices has changed private therapy. Nonetheless, the ethical use of individual information is crucial for a patient-centric future in the healthcare market. The collection, storage space, and also evaluation of individual information are important for technology and personalized healthcare circulation. To produce a patient-centric future, durable information administration structures concentrating on personal privacy, consent in addition to protection are essential. Openness to details and clear standards for information usage as well as sharing are essential for constructing a count on electronic health options. Honest detail approaches have to additionally resolve variants coupled with choices in information collection and evaluation. By concentrating on personal privacy together with resolving information use variants, the healthcare market can harness the transformative capacity of development while advertising patient-centered therapy principles.

REFERENCES

[1]   P. Gupta, U. Ghugar, C.V. Vardhan, R. Nayak, and C.S. Rajpoot, "A novel design and development model for people counting in a closed environment with machine learning approach", *OPJU International Technology Conference on Emerging Technologies for Sustainable Development (OTCON),* pp. 1-5, 2023.
        [http://dx.doi.org/10.1109/OTCON56053.2023.10114007]

[2]   K.Y. Ong, P.S.S. Lee, and E.S. Lee, "Patient-centred and not disease-focused: A review of guidelines and multimorbidity", *Singapore Med. J.,* vol. 61, no. 12, pp. 584-590, 2020.
        [http://dx.doi.org/10.11622/smedj.2019109] [PMID: 31489434]

[3]   A. Seppälä, P. Nykänen, and P. Ruotsalainen, "Development of personal wellness information model for pervasive healthcare", *J. Comput. Netw. Commun.,* vol. 2012, pp. 1-10, 2012.
        [http://dx.doi.org/10.1155/2012/596749]

[4]   M. Kovacova, F. Kevicky, and G.H. Popescu, "Generative artificial intelligence-driven healthcare systems in patient record analysis, in disease diagnosis and monitoring, and in customized treatment plans", *Contemp. Read. Law Soc. Justice,* vol. 15, p. 1, 2023.

[5]   L. Chu, A. G. Shah, D. Rouholiman, S. Riggare, and J. G. Gamble, "Patient-centric strategies in digital health." *Digital Health: Scaling Healthcare to the World*, 43-54, 2018.,
        [http://dx.doi.org/10.1007/978-3-319-61446-5_4]

[6]   W.A. Wood, A.V. Bennett, and E. Basch, "Emerging uses of patient generated health data in clinical research", *Mol. Oncol.,* vol. 9, no. 5, pp. 1018-1024, 2015.
        [http://dx.doi.org/10.1016/j.molonc.2014.08.006] [PMID: 25248998]

[7]   R. Rahul, S. Bommareddy, Monika, J. A. Khan, and R. Anand, "A review on healthcare data privacy and security" *Netw. Technol. Smart Healthc.*, 165-187, 2022.,

[8]   K. Campbell, and K. Parsi, "A new age of patient transparency: An organizational framework for informed consent", *J. Law Med. Ethics,* vol. 45, no. 1, pp. 60-65, 2017.
        [http://dx.doi.org/10.1177/1073110517703100] [PMID: 28661284]

[9]   J. Reddy, N. Elsayed, Z. ElSayed, and M. Ozer, "A review on data breaches in healthcare security systems", *Int. J. Comput. Appl.,* vol. 184, no. 45, pp. 1-7, 2023.
        [http://dx.doi.org/10.5120/ijca2023922333]

[10]   M. Fadler, and C. Legner, "Data ownership revisited: Clarifying data accountabilities in times of big data and analytics", *J. Bus. Anal.,* vol. 5, no. 1, pp. 123-139, 2022.
        [http://dx.doi.org/10.1080/2573234X.2021.1945961]

[11]   F. Sanmarchi, F. Toscano, M. Fattorini, A. Bucci, and D. Golinelli, "Distributed solutions for a reliable data-driven transformation of healthcare management and research", *Front. Public Health,* vol. 9, p. 710462, 2021.
        [http://dx.doi.org/10.3389/fpubh.2021.710462] [PMID: 34307291]

[12]   D.S. Char, M.D. Abràmoff, and C. Feudtner, "Identifying ethical considerations for machine learning healthcare applications", *Am. J. Bioeth.,* vol. 20, no. 11, pp. 7-17, 2020.
        [http://dx.doi.org/10.1080/15265161.2020.1819469] [PMID: 33103967]

[13]   R. McClelland and C.M. Harper, "Information privacy in healthcare—The vital role of informed consent", *Eur. J. Health Law* 1.aop, 1-12, 2022.,
[http://dx.doi.org/10.1163/15718093-bja10097]

[14]   J. van de Hoven, "Towards a digital ecosystem of trust: Ethical, legal and societal implications", *Opinio Juris In Comparatione,* vol. 1/2021, pp. 131-156, 2021.

[15]   M.R. Andrus, and M.T. Roth, "Health literacy: A review", *Pharmacotherapy,* vol. 22, no. 3, pp. 282-302, 2002.
[http://dx.doi.org/10.1592/phco.22.5.282.33191] [PMID: 11898888]

[16]   M. Waller, and C. Stotler, "Telemedicine: A Primer", *Curr. Allergy Asthma Rep.,* vol. 18, no. 10, p. 54, 2018.
[http://dx.doi.org/10.1007/s11882-018-0808-4] [PMID: 30145709]

[17]   R.M. Epstein, and R.L. Street Jr, "Shared mind: Communication, decision making, and autonomy in serious illness", *Ann. Fam. Med.,* vol. 9, no. 5, pp. 454-461, 2011.
[http://dx.doi.org/10.1370/afm.1301] [PMID: 21911765]

[18]   J.A. Naslund, K.A. Aschbrenner, L.A. Marsch, and S.J. Bartels, "The future of mental health care: Peer-to-peer support and social media", *Epidemiol. Psychiatr. Sci.,* vol. 25, no. 2, pp. 113-122, 2016.
[http://dx.doi.org/10.1017/S2045796015001067] [PMID: 26744309]

[19]   D. Pacquiao, "Advocacy and empowerment of individuals, families and communities", *Global Appl. Cultur. Compet. Health Care: Guidel. Pract.,* 239-253, 2018.,
[http://dx.doi.org/10.1007/978-3-319-69332-3_27]

[20]   J. G. McNutt, "Data justice and international development: An ethical imperative for policy and community practice.", *The Routledge Handbook of Social Work Ethics and Values.* Routledge, 280-286, 2019.,

[21]   A. Siena, S. Ingolfo and J. Mylopoulos, "Establishing regulatory compliance for information system requirements: An experience report from the health care domain", *29th International Conference on Conceptual Modeling,* 2010.
[http://dx.doi.org/10.1007/978-3-642-16373-9_7]

[22]   R. Nayak, A.K. Nanda, P. Awasthi, and L. Kumar, "Multiple private keys with NTRU cryptosystem", *Int. J. Res. Comput. Commun. Technol.,* vol. 4, no. 3, pp. 250-255, 2015.

[23]   R. Sorensen, and R. Iedema, "Redefining accountability in health care: Managing the plurality of medical interests", *Health,* vol. 12, no. 1, pp. 87-106, 2008.
[http://dx.doi.org/10.1177/1363459307083699] [PMID: 18073248]

[24]   C. Brall, P. Schröder-Bäck, and E. Maeckelberghe, "Ethical aspects of digital health from a justice point of view", *Eur. J. Public Health,* vol. 29, suppl. Suppl. 3, pp. 18-22, 2019.
[http://dx.doi.org/10.1093/eurpub/ckz167] [PMID: 31738439]

SUBJECT INDEX

www.ingramcontent.com/pod-product-compliance
Lightning Source LLC
Chambersburg PA
CBHW050832220326
41598CB00006B/360